Texts in Computer Science

Series Editors

Orit Hazzan ⓘ, Faculty of Education in Technology and Science, Technion—Israel Institute of Technology, Haifa, Israel

Frank Maurer, Department of Computer Science, University of Calgary, Calgary, Canada

Titles in this series now included in the Thomson Reuters Book Citation Index!

'Texts in Computer Science' (TCS) delivers high-quality instructional content for undergraduates and graduates in all areas of computing and information science, including core theoretical/foundational as well as advanced applied topics. TCS books should be reasonably self-contained and aim to provide students with modern and clear accounts of topics ranging across the computing curriculum. As a result, the books are ideal for semester courses or for individual self-study in cases where people need to expand their knowledge. All texts are authored by established experts in their fields, reviewed internally and by the series editors, and provide numerous examples, problems, and other pedagogical tools; many contain fully worked solutions.

The TCS series is comprised of high-quality, self-contained books that have broad and comprehensive coverage and are generally in hardback format and sometimes contain color. For undergraduate textbooks that are likely to be more brief and modular in their approach, Springer offers the flexibly designed *Undergraduate Topics in Computer Science* series, to which we refer potential authors.

"Writing a book is a horrible, exhausting struggle, like a long bout of some painful illness. One should never undertake such a thing if one were not driven by some demon whom one can neither resist nor understand." George Orwell, "Why I Write", 1946.

"And further, by these, my son, be admonished: of making many books there is no end; and much study is a weariness of the flesh." Ecclesiastes 12:12, King James Version.

"A processor is dead without animating code. A user has to have software to establish contact with the hardware." Language Saved in Calligraphy Lesson: the Collected Stories, Mikhail Shishkin.

For Irene

George Delic

Guide to Numerical Algorithm Design and Development

Including Legacy Examples from Fortran and MathCAD in High Precision

George Delic
University Fellow
University of Wollongong
Wollongong, NSW, Australia

ISSN 1868-0941 ISSN 1868-095X (electronic)
Texts in Computer Science
ISBN 978-3-031-90177-5 ISBN 978-3-031-90178-2 (eBook)
https://doi.org/10.1007/978-3-031-90178-2

© The Editor(s) (if applicable) and The Author(s), under exclusive license to Springer Nature Switzerland AG 2026

This work is subject to copyright. All rights are solely and exclusively licensed by the Publisher, whether the whole or part of the material is concerned, specifically the rights of translation, reprinting, reuse of illustrations, recitation, broadcasting, reproduction on microfilms or in any other physical way, and transmission or information storage and retrieval, electronic adaptation, computer software, or by similar or dissimilar methodology now known or hereafter developed.
The use of general descriptive names, registered names, trademarks, service marks, etc. in this publication does not imply, even in the absence of a specific statement, that such names are exempt from the relevant protective laws and regulations and therefore free for general use.
The publisher, the authors and the editors are safe to assume that the advice and information in this book are believed to be true and accurate at the date of publication. Neither the publisher nor the authors or the editors give a warranty, expressed or implied, with respect to the material contained herein or for any errors or omissions that may have been made. The publisher remains neutral with regard to jurisdictional claims in published maps and institutional affiliations.

This Springer imprint is published by the registered company Springer Nature Switzerland AG
The registered company address is: Gewerbestrasse 11, 6330 Cham, Switzerland

If disposing of this product, please recycle the paper.

Foreword

George Delic is well-known in high performance computing circles for his extensive work on the development of programs designed to provide numerical solutions of complicated problems in Theoretical Physics and related fields. He has worked in four continents—Australia, Germany, Africa and America—seeking ways to tame the computer to his needs. In his efforts he has developed tools for fellow theoreticians to employ in their explorations of complex problems. This monograph brings together the results of many decades of innovation so that the scientific community can share in the fruits of his labours. I commend this text for the serious perusal of explorers of the unknown, not only in Theoretical Physics but also in the diverse fields which are amenable to treatment with the codes presented here.

P. G. L. Leach
Professor of Mathematics Emeritus
University of KwaZulu-Natal
Durban, Republic of South Africa

Preface

The author's coding experience is predominantly with Fortran and developed throughout his multi-continent, multi-career path, as knowledge and skill accumulated in algorithm development and coding. This began in 1965 while an undergraduate, then next as a Ph.D. student and later with experience on High Performance Computing (HPC) platforms using vector and parallel computers. During this time the Fortran language standard has evolved over 70 years after its inception, it remains a bulwark of large scale science and engineering numerical models when high performance is required.

There are several motivations for writing this book. The first among them is to collect the algorithms in Fortran code developed by the author over his career as a computational physicist working in the area of theoretical nuclear physics. A second motivation is the wish to pass on to a new generation of computational scientists and engineers, some documentation and description of several related topics in numerical analysis that may be of value to them in saving development time for algorithms, as they progress in modeling larger problems. A third motivation is to support the algorithms presented here by collating relevant material from several courses taught by the author in universities in the USA and elsewhere. The mathematical details are presented as background material to the algorithms in an effort to make the book as self contained as possible, without overloading the reader with too much detail. For those wishing to pursue the issues in greater detail there is an ample bibliography with citations throughout the text. Taken together it is hoped that this book is a legacy to the next generation of computational specialists active in large scale model development across science and engineering.

For a novice it is worthwhile remembering that the history of algorithms was a path followed by many explorers in the past and their biography is also of interest in following their development of mathematical and numerical science [1, 2]. For this reason, when a result is attributed to a historical discoverer, or researcher, a footnote gives a brief mention.[1]

---

[1] For detailed biographies of mathematicians, follow this link https://mathshistory.st-andrews.ac.uk/Biographies/.

Key to successful implementation of any algorithm is the necessity of three criteria:

- Stability: knowledge of the range of parameters for which the algorithm is known to produce reliable results.
- Accuracy: the extent to which the results are known to be accurate in a quantitative metric.
- Precision: the expected numerical precision of results must be known in advance and controlled in the implementation.

Throughout the exposition these three criteria are examined and demonstrated in detail by examining methods of discovery that produce results of accuracy in many cases that are beyond what is available in the published literature (as is highlighted below in the Section About This Book and How to Use It).

Durham, NC, USA   George Delic

Acknowledgments

The author acknowledges those teachers who mentored him on much of the matter discussed in this book. Professor Austin Keane†, University of Wollongong (UOW), Australia, lectured on classical applied mathematics and introduced his class to Fortran programming in the second year of the undergraduate curriculum. The author's Ph.D. thesis adviser, Dr. B. A. Robson, at the Institute of Advanced Studies, Australian National University (ANU), Australia, set him on the path as a computational physicist. Lastly, Prof. Friedrich Beck†, Institut für Kernphysik, Technische Hochschule Darmstadt, Germany, encouraged the author's independence in his own research in theoretical nuclear physics on his first overseas post-doctoral appointment in Germany.

About This Book and How to Use It

This book is not a course in how to write computer code, instead its purpose is to teach computer algorithm design and the writing of code is left as an exercise for the reader. This book is best used by following the exposition first, then studying the numbered MathCAD[2] examples in each section (where they appear). Next the Fortran code(s), as summarized in Table 2 (below), are shown as examples of implementation and should be compiled and executed after download from git hub.com (see the footnote below). The codes come with Makefiles for a Linux operating system and have been tested using the Intel^{TM} Corporation Fortran compiler.[3] Commentary in the accompanying Fortran code and README text files have more detail, particularly regarding the choice of compiler options and precision of floating point variables. End-users of the book will have to make appropriate adjustments if they prefer to choose another Fortran compiler, such as Portland,[4] now part of the NVIDIA HPC SDK[5] as nvfortran, or GNU.[6] Readers are encouraged to use the MathCAD examples as a basis to develop their own code in HPC languages such as Fortran, C/C++, or an interpretive compiler such as Julia with HPC options (for an overview see [3]). Detailed texts should facilitate such an enterprise and are listed here at either introductory [4–6], or advanced levels [3, 7, 8], all with sample code examples.

A short road-map of the book is as follows.

- Chapter 1 introduces the background to modeling, the number systems used with computers and the limitations of numerical precision. The discussion shows how hardware limits lead to error growth in numerical computation and the importance of monitoring error propagation. Code examples show how the platform may be interrogated using Fortran intrinsic functions to determine the exponentiation range and precision retained for numerical representations. A brief summary of the history of Fortran's development is also presented.

---

[2] www.mathcad.com.
[3] http://www.intel.com.
[4] https://www.pgroup.com/.
[5] https://developer.nvidia.com/hpc-sdk.
[6] https://gcc.gnu.org/fortran/.

- Chapters 2–4 should be studied sequentially and by using the examples to view the numerical approximation methods in action. The discussion covers the choice of basis sets to approximate both continuous functions on a finite interval and cases when only discrete data values are available for approximation. The importance of discovering how error is related to approximation methods is developed and how a "best approximation" may be quantitatively determined. Metrics for this are covered in detail in Chap. 4 with application in developing algorithms for the general approximation problem. Several Fortran code samples demonstrate the details.
- Chapter 5 defines vectors and matrices and uses them as examples of operator equations that drive approximation and error analysis. Several methods of solving sets of linear equations are explained and eigenvalue operators are included. Both dense and sparse matrix algebra is discussed to study stability and convergence criteria.
- Chapter 6 discusses root finding methods for functions where a numerical algorithm is the only option. Three methods are compared and roots of polynomials and higher transcendental functions serve as suitable candidates for code examples. Some of the root values found are for use in the subsequent chapters.
- Chapters 7 and 8 introduce numerical methods for the evaluation of integrals of functions in both one- and two-dimensional (or higher) cases. For the former case, fixed step and polynomial interpolation quadrature methods are discussed for both monotonic and oscillating weight functions. The two-dimensional case shows the advantage of using minimal point quadrature formulas over product formulas of either fixed point, Gauss type formulas, or Monte Carlo methods.
- Chapter 9 discusses the numerical solution of first and second order ordinary differential equations using function approximation methods (introduced in Chap. 2) in matrix operator equation methods from Chap. 5. Error metrics developed in Chap. 4 are applied to study convergence and stability in solutions. Applications include the diffusion and wave equations.
- Chapter 10 introduces two direct search optimization algorithms for finding the global minima of functions in multidimensional spaces.

Copyright, Terms of Use and Trademarks

1. All rights reserved. Without limiting the rights under copyright, no part of this document may be reproduced, stored in or introduced into a retrieval system, or transmitted in any form or by any means (electronic, mechanical, photocopying, recording, or otherwise), or for any purpose, without the written permission of the publisher.
2. This document has adapted some material from other sources as cited in the first column of Table 1 where the reference to a specific section or chapter is denoted by parentheses {..} and (..), respectively. The location in this book where such examples appear is shown in the second column of Table 1 (with similar parentheses notation). The third column shows copyright ownership. Due diligence by this author for such material is hereby acknowledged and is reproduced with the permission of the authors and/or publishers where such information is publicly available. Regrettably, with a time span reaching back over half a century, some of the original (former) publishers are no longer extant or (current) copyright has been acquired by other publishers (as shown in the third column of Table 1). In a few cases flagged with "(?)" copyright ownership is not traceable (or authors are deceased). For example a search at https://marketplace.copyright.com/rs-ui-web/mp on the citation [9] returned a message recommending contact with the publisher or author (deceased in this case). Prentice-Hall was acquired by Gulf + Western in 1984 as part of their publishing division Simon & Shuster. In 1998 Pearson, https://www.pearson.com (the original owner of Prentice-Hall), purchased the education division of Simon & Schuster, but this publication is not found under their imprint.
3. The author retains the copyright to modifications of code examples taken from other sources, or source code developed by the author for the first time whether published or not. The source code distributed with this book, is included with the GNU General Public License[1] unless otherwise stated in the code when it appears in publications by this author or by others. A summary is shown in Table 2 and complete examples may be downloaded from github.[2] Such code

---

[1] http://www.gnu.org/licenses/gpl.html.
[2] https://github.com/hiperism/legacy_book_FTN_examples.

Table 1 Permissions

Citation	Location	Copyright current/former
[10]	{1.9}	O'Reilly and Associates
[11] p. 43	{2.4} (2.58)	Elsevier/Academic Press
[12] {9.3.6} (Chap. VIII)	{2.6} (2.110)	Elsevier/Academic Press
[13]	{2.6} (2.118)–(2.121)	Elsevier
[14]	{2.7} (2.128)–(2.137)	Elsevier
[15] pp. 73–75 (Chap. V)	{2.8} (2.139)	Academic Press (?)
[16]	{2.9}	Elsevier
[17] {1.2}–{1.6}	{3.9, 3.10} (2.77)–(2.102)	John Wiley and Son (?)
[18] pp. 32–35	{5.1}	Springer-Verlag
[19]	{5.4}	SIAM[a]
[20]	{5.5}	Elsevier
[21]	{7.5}	Elsevier
[22]	{8.2.1} (8.2)–(8.3)	Blaisdell Publishing/(?)
[23] (Chaps. 2, 4, 5)	{9.3.3}	Elsevier/Academic Press[b]
[24]	{9.3}	Elsevier
[9]	{9.3.5}	Prentice-Hall Inc./(?)
[25]	{9.4}	AIP[c]
[26] (Chaps. 5, 6)	{9.4.6, 9.5, 9.8}	Springer-Nature/Reidel
[27] {3.3}	{9.6}	Springer-Verlag
[28] {II.13}	{9.6}	Springer-Verlag
[29] {1.5(III)}	{10.3.1}	English Universities Press/(?)
[30] {2}	{10.3.2}	Oxford University Press

[a] Society for Industrial and Applied Mathematics.
[b] Elsevier is unable to confirm they hold the copyright but the author gratefully acknowledges John P. Seinfelds's permission for "free use of material from this book".
[c] Reproduced with permission of American Institute of Physics Publishing.

may be copied if the header and README file information in each case is contained in the copy. The use of all source code described in this document is at the end user's own risk.

4. Named companies, products and services, referenced herein are the trademarks of their respective owners. Discussion of products by the author in this document implies no endorsement of these products. Specific examples are Fortran compilers from Intel[3] and the Portland Group.[4] Examples that accompany the text have been prepared using PTC MathCAD Prime™, a trademark of PTC[5]

---

[3] www.intel.com.
[4] www.pgroup.com and www.nvidia.com.
[5] www.mathcad.com.

Table 2 Index of Fortran examples

Example name	Author	Location	Citation where published
KERRIGAN	James F. Kerrigan	1.9	Adapted from [10]
JLRCHB	George Delic	2.6	[13]
PLMCHB	George Delic	2.7	[14]
DLRCHB	George Delic	2.8	Unpublished report and code
DLRLUK	Yudell L. Luke	2.8	Adapted from [15]
CCFS	George Delic	2.9	[16]
STIR	George Delic	3.11	Unpublished report and code
FSPARSE	George Delic	5.4	Adapted from [19]
PERM	George Delic	5.5	[20]
PLZEROES	George Delic	6.9, 7.5	Unpublished report and code
GAUSSLEG	George Delic	6.9, 7.4.1	[21]
RKPC	George Delic	9.3.3, 9.3.4	[24]
GEAR	C. William Gear	9.3.5	Adapted from [9]
HJ	George Delic	10.2	Unpublished report and code
DSC	George Delic	10.3	Unpublished report and code

and purchase of these products in following the examples is not required to comprehend them. All MathCAD examples template may be downloaded from github.[6] The use of these products is at the users own risk.

References

1. Mitchell JDS, Tootill E (1975) Chambers biographical encyclopedia of scientists. W and R Chambers, Edinburgh, UK
2. Arianrhod R (2024) Vector: a surprising story of space, time, and mathematical transformation. University of Chicago Press, Chicago, USA
3. Sengupta A (2019) Julia high performance, 2nd ed Packt Publishing Ltd., Birmingham, UK
4. Balbaert I (2015) Getting started with Julia Programming. Packt Publishing Ltd., Birmingham, UK
5. Lobianco A (2019) Julia quick syntax reference. Springer Science+Business Media, New York, NY
6. Lauwens B, Downey AB (2019) Think Julia how to think like a computer scientist. O'Reilly and Associates, Sebastopol, CA
7. Kwong T (2020) Hands-on design patterns and best practices with Julia. Packt Publishing Ltd., Birmingham, UK

---

[6] https://github.com/hiperism/legacy_book_MCAD_examples.

8. Darve E, Wootters M (2021) Numerical linear algebra with Julia. Society for industrial and applied mathematics, Philadelphia, PA
9. Gear CW (1971) Numerical initial value problems in ordinary differential equations. Prentice-Hall, Inc., Englewood Cliffs, NJ
10. Kerrigan JF (1993) Migrating to Fortran 90. O'Reilly and Associates, Sebastopol, CA
11. Luke YL (1975) Mathematical functions and their approximations. Academic Press, New York
12. Luke YL (1969) The special functions and their approximation, vols I and II. Academic Press, New York
13. Delic G (1979) Chebyshev series for the spherical Bessel function jl(r). Comput Phys Commun 18:73–86
14. Delic G (1979) Chebyshev expansion of the associated legendre polynomial plm(x). Comput Phys Commun 18:63–71
15. Luke YL (1977) Algorithms for the computation of mathematical functions. Academic Press, New York
16. Delic G, Malherbe SM (1988) Subroutines for convolution sums of Fourier and Chebyshev series. Comput Phys Commun 48:305–312
17. Herriot JG (1963) Methods of mathematical analysis and computation. John Wiley and Sons, New York, NY
18. Flügge S (1984) Practical quantum mechanics. Springer-Verlag, Berlin
19. Davis TA (2006) Direct methods for sparse linear systems. Society for Industrial and Applied Mathematics, Philadelphia, PA
20. Delic G, Cash GG (2000) The permanent of 0,1 matrices and Kallman's algorithm. Comput Phys Commun 124:315–329
21. Delic G (1974) The legendre series and a quadrature formula for its coefficients. J Comput Phys 14:254–268
22. Davis PJ, Rabinowitz P (1967) Numerical integration. Blaisdell Publishing Company, Waltham, MA
23. Lapidus L, Seinfeld JH (1971) Numerical solution of ordinary differential equations. Academic Press, New York, NY
24. Delic G, Malherbe SM (1988) Subroutines for the integration of systems of first order ode's. Comput Phys Commun 48:293–304
25. Delic G (1987) Spectral function methods for nonlinear diffusion equations. J Math Phys 28:39–59
26. Rektorys K (1982) The method of discretization in time. D. Reidel Publishing Company, Dordrecht, Holland
27. Nettel S (1992) Wave physics: oscillations-solitons-chaos. Springer-Verlag, Berlin
28. Hairer E, Norsett SP, Wanner G (1987) Solving ordinary differential equations I nonstiff problems. Springer-Verlag, Berlin
29. Dixon LCW (1972) Nonlinear optimization. The English Universities Press Ltd., London, UK
30. Powell MJD (1968) On the calculation of orthogonal vectors. Comput J 11:302–304
31. Delic G (2019) A thread parallel sparse matrix chemistry algorithm for the community multiscale air quality model. Mod Environ Sci Eng 5:775–791

Contents

1 Number Systems and Machine Representation 1
  1.1 Introduction 1
  1.2 Computer Simulation and Models 1
  1.3 The Decimal Number System 2
  1.4 The Binary Number System 3
  1.5 The Octal and Hexadecimal Number Systems 5
  1.6 Change of Base in a Numerical Model 6
  1.7 Machine Representation of Numbers 7
  1.8 Computer Error and Loss of Significance 8
  1.9 Fortran 10
  1.10 Exercises 13
  1.11 Programming Problems 13
  References 13

2 Function Approximation and Error 15
  2.1 Function Approximation by Recurrence 15
  2.2 Approximation with a Set of Basis Functions 17
  2.3 Monomial Basis Functions 18
  2.4 Polynomial Basis Functions 22
  2.5 Trigonometric Basis Functions 27
  2.6 Application to Bessel Functions 30
  2.7 Application to the Associated Legendre Polynomial 35
  2.8 Application to the Parabolic Cylinder Function 38
  2.9 Convolution of Series Products 41
  2.10 Types of Convergence 44
  2.11 Differentiation and Integration of Monomial
       and Polynomial Series 47
  2.12 Exercises 49
  2.13 Programming Problems 50
  References 50

3 Interpolation of Discrete Data 53
  3.1 Definition of the Interpolating Polynomial 53
  3.2 The Newton Interpolating Polynomial 55

xxiii

3.3	Divided Differences and Newton Interpolation	57
3.4	The Lagrange Interpolating Polynomial	59
3.5	Error in Polynomial Interpolation	61
3.6	The Derivative of the Interpolating Polynomial	62
3.7	Accelerating Convergence for Derivatives	63
3.8	Finite Differences	65
3.9	The Newton Polynomial Revisited	67
3.10	Stirling's Central Difference Formula	70
3.11	Application of Stirling's Formula	72
3.12	Error in Stirling's Formula	79
3.13	Exercises	82
3.14	Programming Problems	83
	References	83

4 Function Approximation ... 85
- 4.1 Measuring the Quality of Approximation ... 85
- 4.2 Metric Spaces and Distance ... 85
- 4.3 Normed Spaces and Size ... 88
- 4.4 Examples of Normed Spaces ... 89
- 4.5 Hilbert Space ... 91
- 4.6 The Orthonormal Basis in Hilbert Space ... 93
- 4.7 The General Approximation Problem ... 96
- 4.8 The L^2 Error and Truncated Series Expansions ... 98
- 4.9 Applications ... 102
- 4.10 Exercises ... 104
- 4.11 Programming Problems ... 104
- References ... 104

5 Operator Equations and Notation ... 107
- 5.1 Operator Equations ... 107
- 5.2 Vectors ... 109
- 5.3 Matrices ... 110
- 5.4 Sparse Matrix Storage ... 114
- 5.5 Determinants ... 116
- 5.6 Matrix Operator Equations ... 118
- 5.7 Matrix Solution Methods ... 120
  - 5.7.1 Gaussian Elimination ... 121
  - 5.7.2 Partial Pivoting ... 124
  - 5.7.3 The Inverse Matrix ... 125
  - 5.7.4 The LU Decomposition ... 126
  - 5.7.5 Crout and Cholesky Decomposition ... 129
  - 5.7.6 Tridiagonal Matrices ... 130
  - 5.7.7 Gauss-Jordan Elimination ... 132
- 5.8 Matrix Norms, Condition Number and Stability ... 135
- 5.9 Iterative Solution Methods ... 140
- 5.10 Convergence Criteria ... 143

	5.11	Relaxation	146
	5.12	Sparse Matrices: Case Study	147
	5.13	Exercises	148
	5.14	Programming Problems	150
	References		151

6 Finding Roots of Functions ... 153
- 6.1 Definition and Examples ... 153
- 6.2 Numerical Root Finding Methods ... 154
- 6.3 The Bisection Method ... 154
- 6.4 The Newton Method ... 156
- 6.5 The Secant Method ... 159
- 6.6 Comparison of Methods ... 160
- 6.7 Finding Roots of Ill-Posed Problems ... 161
- 6.8 Roots of the Spherical Bessel Function: Case Study ... 162
- 6.9 Roots of the Legendre Polynomial: Case Study ... 162
- 6.10 Exercises ... 163
- 6.11 Programming Problems ... 163
- References ... 164

7 One-Dimensional Numerical Integration ... 165
- 7.1 Definition of the Integral ... 165
- 7.2 The Riemann Integral ... 166
- 7.3 Fixed Step Formulas ... 168
  - 7.3.1 The Trapezoidal Rule ... 168
  - 7.3.2 Simpson's Rule ... 172
  - 7.3.3 Higher Order Rules ... 175
- 7.4 Gaussian Quadrature ... 176
  - 7.4.1 Gauss-Legendre Formulas ... 176
  - 7.4.2 Chebyshev Formulas ... 182
- 7.5 Special Case of an Oscillating Weight Function ... 183
- 7.6 Exercises ... 185
- 7.7 Programming Problems ... 187
- References ... 187

8 Two-Dimensional Numerical Integration ... 189
- 8.1 Definition of the Integral ... 189
- 8.2 Product Formulas ... 190
  - 8.2.1 Simpson's Rule ... 190
  - 8.2.2 Gauss-Legendre Quadrature ... 191
- 8.3 Minimal Point Formulas ... 191
- 8.4 The Monte Carlo Method ... 192
- 8.5 Exercises ... 194
- 8.6 Programming Problems ... 194
- References ... 195

9 Numerical Solution of Ordinary Differential Equations ... 197
- 9.1 Background ... 197
- 9.2 Definitions ... 198
- 9.3 Numerical Solution Methods on a Physical Grid ... 198
  - 9.3.1 Euler's Method ... 199
  - 9.3.2 Multi-step Methods ... 199
  - 9.3.3 Runge-Kutta ... 200
  - 9.3.4 Predictor-Corrector ... 204
  - 9.3.5 Stiff ODEs ... 205
- 9.4 Spectral Function Methods ... 208
  - 9.4.1 Spectral Functions ... 208
  - 9.4.2 Choice of Basis ... 208
  - 9.4.3 Incorporating Boundary Conditions ... 209
  - 9.4.4 Orthonormalization ... 211
  - 9.4.5 The Second-Order Spatial Derivative ... 212
  - 9.4.6 Spectral Mappings ... 214
  - 9.4.7 A Case Study of the Non-linear Diffusion Equation ... 216
- 9.5 Double Spectral Function Method ... 217
  - 9.5.1 The Non-linear Diffusion Equation Solved by Time Stepping ... 218
  - 9.5.2 The Non-linear Diffusion Equation Solved by Recurrence ... 220
- 9.6 Second-Order ODEs ... 224
- 9.7 Exercises ... 228
- 9.8 Programming Problems ... 228
- References ... 230

10 Direct Search Optimization Methods ... 233
- 10.1 Why Direct Search Methods? ... 233
- 10.2 Hooke and Jeeves Algorithm ... 234
- 10.3 Davies, Swann, and Campey Algorithm ... 236
  - 10.3.1 The Fibonacci Search ... 236
  - 10.3.2 Gram-Schmidt Orthonormalization ... 237
- 10.4 Comparison of the Algorithms ... 238
- 10.5 Programming Problems ... 239
- References ... 240

Index ... 241

About the Author

George Delic attained his Bachelor of Science degree in 1967 at the Wollongong University College, now the University of Wollongong (UOW), where he was elected University Fellow in 2022. His experience included, not only the theory of many of the functions and methods described in this book, but also understanding and exploring their properties when approximated. Ph.D. studies followed at the Institute of Advanced Studies, Australian National University (ANU), in Canberra, with graduation in early 1971. From there, post-doctoral appointments in Germany and Lawrence Berkeley National Laboratory in the USA followed. The publication record shows the growth of code development for numerical algorithms. A later appointment was in the Physics Department at the University of the Witwatersrand (WITS), Johannesburg, South Africa. In addition to research, student supervision and more work with numerical approximation followed with lecture course presentation on functional analysis and scattering theory. After this background, a return to the USA followed with appointment at the Ohio Supercomputer Center. This time included teaching a numerical analysis course in the Computer and Information Science Department at the Ohio State University (OSU), Columbus, Ohio, from which some material in this book originates. A later appointment was as a contractor to the US Environmental Protection Agency, supporting EPA scientists in porting their environmental models to High Performance Computing (HPC) platforms. As a former contractor for NASA and the EPA, the author is currently engaged as a private HPC consultant and implementing recent developments in parallel algorithms for sparse matrix solvers for environmental and climate models [31].

Number Systems and Machine Representation

1.1 Introduction

This chapter sets the scene for the remainder of the book with a review of four disparate, but key components (for later discussion) and examples for numerical computing:

- the relationship between simulation, models and computers,
- numerical models and machine representation,
- computer error and loss of precision,
- the Fortran Language.

The first topic, discussed in the next section, sets the context for the book. Other sections define elementary resources used in this discussion. The inherent unity of these elements in the computing environment are a fundamental part of reliable algorithm development throughout this book.

1.2 Computer Simulation and Models

New computational resources in High-Performance Computing (HPC), usually consist of clusters of independent nodes, each with shared memory, with each node having multiple cores and each core having vector processor arithmetic units. Today scientific applications developers are formulating parallel numerical algorithms with

---

Supplementary Information The online version contains supplementary material available at https://doi.org/10.1007/978-3-031-90178-2_1

flexible data structures using advanced computer languages for this purpose. Scientists and engineers often discuss the role of computer simulation and identify common components where HPC makes important contributions using criteria such as:

- Prediction and understanding
- Trade-offs between complexity and utility
- Modeling cross-component links
- Current and future model computation
- Model support systems
- Porting code across architectures and performance issues.

At the core of model simulations is the need for robust and efficient numerical algorithms. A collection of algorithms developed by the author is described in sufficient detail in separate chapters, with examples written in Fortran code in some cases. The Fortran examples are prototypes developed as components of larger numerical models over a period of five decades.

Specific issues to be addressed in an integrated approach to model development include:

- PORTABILITY: transparent implementation of models on a range of computational platforms from desktop personal computers, to clusters of servers,
- SCALABILITY: applicability of the models to scaling with system complexity and dimension as resources advance,
- INTERFACE: interfacing of the models with data base systems for input of data and graphical display of simulation results and
- ADVANCED TECHNOLOGY: inclusion of recent advances in hardware and software.

The need to improve computational algorithms, while insuring that numerical precision is preserved, adds to the overall complexity of the model and the programming effort required as models demand large computational resources extending to the limits of available computing capacity.

To begin, the following sections describe how numbers are represented in their machine interface and how computer error is a source of loss in numerical significance. This chapter concludes with a brief survey of the Fortran language's history and evolution over 70 years.

1.3 The Decimal Number System

The decimal number system is the most familiar one because it is based on the decade and, as such, is in common use. Consider a simple decimal number such as 5638, which could be represented as

$$5638 = 8 \times 10^0 + 3 \times 10^1 + 6 \times 10^2 + 5 \times 10^3.$$

This representation shows two important features: (a) the occurrence of the base 10 and (b) the exponent of this base which determines the order of the digits from right to left as the exponent decreases. This example showed the case of an integer number, but consider also the case of a fraction such as 0.4792

$$0.4792 = 4 \times 10^{-1} + 7 \times 10^{-2} + 9 \times 10^{-3} + 2 \times 10^{-4},$$

where the base is unchanged but the exponent now ranges over decreasing negative values which also determine the digit order. It follows that any decimal number has the general representation

$$c_n c_{n-1} \cdots c_1 c_0 . d_1 d_2 d_3 \cdots = \sum_{k=0}^{n} c_k 10^k + \sum_{k=1}^{\infty} d_k 10^{-k}, \quad (1.1)$$

which shows that, in general, the integer part has a terminating string of digits but the fractional part may have an infinite string. The decimal system requires that each coefficient c_k or d_k takes a value in the range between 0 and 9. However, there is nothing unique about the choice of 10 as a base to represent a number. Introducing an arbitrary base B, (1.1) can be generalized to represent a number with respect to this base in the numerical model

$$\{c_n c_{n-1} \cdots c_1 c_0 . d_1 d_2 d_3 \cdots\}_B = \sum_{k=0}^{n} c_k B^k + \sum_{k=1}^{\infty} d_k B^{-k}. \quad (1.2)$$

The notation $\{\}_B$ is introduced when there may be ambiguity as to which base is being used. The only complication which may arise in the use of (1.2) occurs if $B > 10$ in which case a special notation has to be introduced for coefficient values larger than 9, namely, those for 10 through $B - 1$. The choice of $B = 10$ is convenient for human use, but other choices, more common for machine use, are discussed in the following two sections.

EXAMPLE 1.1

1.4 The Binary Number System

The numerical model with the simplest base is the binary system where $B = 2$. This has the oldest history in digital computing because the coefficients, c_k, d_k, in (1.2) have values of either 0 or 1 only and this was convenient in digital circuits because they could be represented by an off/on switch. Binary numbers are also important in Information Theory [1]. Any number can have different representations and there is always a one-to-one correspondence. Thus, for example, the first few decimal numbers, 0 to 6, are related to the binary representation as follows:

$$0 = 0 \times 2^0 + 0 \times 2^1 + 0 \times 2^2 \equiv 000$$
$$1 = 1 \times 2^0 + 0 \times 2^1 + 0 \times 2^2 \equiv 001$$
$$2 = 0 \times 2^0 + 1 \times 2^1 + 0 \times 2^2 \equiv 010$$
$$3 = 1 \times 2^0 + 1 \times 2^1 + 0 \times 2^2 \equiv 011 \,. \tag{1.3}$$
$$4 = 0 \times 2^0 + 0 \times 2^1 + 1 \times 2^2 \equiv 100$$
$$5 = 1 \times 2^0 + 0 \times 2^1 + 1 \times 2^2 \equiv 101$$
$$6 = 0 \times 2^0 + 1 \times 2^1 + 1 \times 2^2 \equiv 110$$

As an example, the decimal number 5638 has the binary representation

$$5638 = 1 \times 2^1 + 1 \times 2^2 + 1 \times 2^9 + 1 \times 2^{10} + 1 \times 2^{12}, \tag{1.4}$$

with zero coefficients for all other exponents of the base 2. This is commonly written as a binary string in packed format

$$\{5638\}_{10} = \{1011000000110\}_2. \tag{1.5}$$

One method of establishing this equivalence is shown in EXAMPLE 1.2 and consists of continued division of the decimal number by 2. In EXAMPLE 1.3 a somewhat fancier (but more complicated) method is used.

Once a numerical model is chosen then the usual arithmetic operations can also be defined in the specific model. This is not done here, but because of the importance of the packed format in integer programming applications, some elementary logical operations are defined as follows for binary strings A, B, and C.

1. I = POPCNT(A) (population count): the number of non-zero bits in the binary string A,
2. A = B.AND.C (logical AND): binary operation on two binary strings B, C with resultant A that has 1 in locations where both binary operands have a 1,
3. A = B.OR.C (logical OR): binary operation on two binary strings B, C with resultant A that has 1 in locations where either (or both) binary operands has a 1 and
4. A = B.XOR.C (logical exclusive OR): binary operation on two binary strings B, C with resultant A that has 1 in locations where either (but not both) binary operands has a 1.

Most computer languages support the logical operations, but not the POPCNT function. It was available on Cray computer systems as an extension to standard Fortran and allowed the author to use it in speeding up the calculation of the permanent of a matrix (see Sect. 5.5) by orders of magnitude. [2]

As a side note, not only have binary strings been used for representing numbers but also for alphabetical characters. As early as 1855, a "printing telegraph" was invented, but only came into major use when bit-wise encryption was introduced by Gilbert Vernam in 1919 for teletype transmissions. This used the logical XOR operation to compare two strings in an encryption algorithm that became an intense field of study associated with the Colossus computer during World War 2 (see Appendix 12 in [3]).

EXAMPLES 1.2 and 1.3

1.5 The Octal and Hexadecimal Number Systems

The octal and hexadecimal number systems are among the most popular in machine use and employ a base B which is a power of 2. The octal representation has $B = 8$ as base and an example is the integer number $\{35731\}_8$

$$\{35731\}_8 = 1 \times 8^0 + 3 \times 8^1 + 7 \times 8^2 + 5 \times 8^3 + 3 \times 8^4, \tag{1.6}$$

which may also be written in nested form to obtain the decimal value from

$$\{35731\}_8 = ((((3)8 + 5)8 + 7)8 + 3)8 + 1, \tag{1.7}$$
$$\equiv \{15321\}_{10}.$$

For a fraction, the corresponding representation would be

$$\{0.4713\}_8 = 4 \times 8^{-1} + 7 \times 8^{-2} + 1 \times 8^{-3} + 3 \times 8^{-4}, \tag{1.8}$$

which may be written in nested form to obtain the decimal value from

$$\{0.4713\}_8 = 8^{-4}(((4)8 + 7)8 + 1)8 + 3, \tag{1.9}$$
$$\equiv \{0.61206\cdots\}_{10}.$$

The hexadecimal model has 16 as base and therefore requires a notation for coefficient values above 9. Usually, the coefficients in the range 10–15 are assigned letters A through F. An example is the integer number $\{3BAD\}_{16}$

$$\{3BAD\}_{16} = D \times 16^0 + A \times 16^1 + B \times 16^2 + 3 \times 16^3, \tag{1.10}$$

which may also be written in nested form to obtain the decimal value from

$$\{3BAD\}_{16} = (((3)16 + 11)16 + 10)16 + 13, \tag{1.11}$$
$$\equiv \{15277\}_{10}.$$

For a fraction, the representation would be, for example,

$$\{0.C7E\}_{16} = 12 \times 16^{-1} + 7 \times 16^{-2} + 14 \times 16^{-3}, \tag{1.12}$$

which may also be written in nested form to obtain the decimal value from

$$\{0.C7E\}_{16} = 16^{-3}(((12)16 + 7)16 + 14), \tag{1.13}$$
$$\equiv \{0.78076\}_{10}.$$

1.6 Change of Base in a Numerical Model

A change of base B is especially simple if the different bases correspond simply to a different power of 2. Thus, because $8 = 2^3$, a change of base from binary to octal is performed as follows. Note that octal coefficients of any number range from 0 to 7 and the simple linear combination

$$c_0 2^0 + c_1 2^1 + c_2 2^2, \tag{1.14}$$

takes values in this range for each possible triplet of binary coefficients $\{c_0, c_1, c_2\}$. Therefore, in a binary number, each triplet of coefficients is simply replaced by the corresponding octal coefficients. For example using (1.3), the conversion of 101101001.1100101, with the triplet groupings displayed, is

$$\{101\ 101\ 001.110\ 010\ 100\}_2 \equiv \{551.624\}_8, \tag{1.15}$$

where the last fractional triplet has been padded with zeros on the right. Because $16 = 2^4$, the same principle applies for conversion from binary to hexadecimal after the binary equivalent of the hexadecimal coefficients 0 through F have been tabulated and quadruplets of binary coefficients replaced.

When the base conversion is from decimal to one of the other representations, then a different approach is to exploit the general expression (1.2) rewritten in nested form. For an integer, this would be

$$\{c_n c_{n-1} \cdots c_1 c_0\}_B = \sum_{k=0}^{n} c_k B^k,$$
$$= (\cdots (c_n) B + c_{n-1}) B + \cdots c_1) B + c_0, \tag{1.16}$$

and division of a decimal integer by the base B gives a remainder of c_0. Continued division of the quotient by B gives successively, $c_1, \cdots, c_{n-1}, c_n$, as remainders. For a fraction, the analogous result to (1.16) is

$$\{.d_1 d_2 d_3 \cdots .\}_B = \sum_{k=1}^{n} d_k B^{-k},$$
$$= (\cdots (\cdots) B^{-4} + d_3) B^{-3} + d_2) B^{-2} + d_1 B^{-1}, \tag{1.17}$$

and multiplication of a decimal integer by the base B gives an integer value of d_1. Continued multiplication of the fractional remainder by B gives, successively, $d_2, d_3, \cdots$, as the integer parts.

EXAMPLES 1.4 and 1.5

1.7 Machine Representation of Numbers

While floating point numbers may have different machine representations, they are invariably presented in the normalized scientific (floating point) representation as a decimal

$$x = \mp 0.d_1 d_2 d_3 \cdots \times 10^n, \quad (1.18)$$

where $d_1 \neq 0, 0 < d_i < 9, i = 1, 2, 3, \cdots$, and the integer exponent is either positive, negative, or zero. The representation consists of the sign, the mantissa, $d_1 d_2 d_3 \cdots$ and the exponent. The representation is said to be normalized because of the requirement that $0.1 \leq 0.d_1 d_2 d_3 \cdots < 1$. In the case of a normalized binary representation, the mantissa would require the fraction to be in the range $0.5 \leq 0.d_1 d_2 d_3 \cdots < 1$ with a base 2.

In machine representation, each binary coefficient is called a bit. A set of bits is a byte with the actual number of bits in a byte dependent on the machine implementation. The number of bits (or bytes) required to represent a floating point number determines the word length of a machine. As an example, in the personal computer (PC) based on the Intel i8086 processor, the length of the binary word used to store floating point numbers is 32 bits and this is subdivided as 1 bit for the sign, 8 bits for the exponent, and the remaining 23 bits for the mantissa. For an integer, the word length is 16 bits of which 1 was used for the sign and the remaining 15 for the binary coefficients c_i. Therefore, the decimal value of the largest integer, in this case, is

$$\sum_{k=0}^{14} 2^k = 32767. \quad (1.19)$$

Because of the limited number of bits used for the exponent, the range from smallest to largest numbers on a i8086 PC is limited to an approximate range of 10^{-39} to 10^{+38}. New generations of commodity personal computers have a 64-bit word length, that is also more typical of former mainframe computers.

In IBM mainframe computers of the past, an integer word had 4 bytes, or 32 bits, so that in this case the largest positive integer is $2^{32-1} = 2147483647$. On this generation of mainframe computer, a floating point number is stored as either a single, double, or quadruple precision constant. In single precision, the word length is 32 bits with 1 bit for the sign, 7 for the exponent, and 24 bits for the mantissa. However, while the exponent is stored as a binary number in the machine, the mantissa is stored as six hexadecimal coefficients of 4 bits each. The corresponding floating point numbers range from approximately 10^{-79} to 10^{+75}. On the other hand, Cray Research Incorporated Supercomputers[1] of the past, had a word length of 64 bits in single precision and this had 1 bit for the sign, 15 bits for the exponent, and 48 bits for the mantissa. In such a case, an integer may have either a 46 or 64-bit word with 2^{46} or 2^{63}, as the respective largest positive integers. The corresponding floating point numbers ranged from approximately 10^{-2467} to 10^{+2465}. The Cray architecture had

---

[1] https://en.wikipedia.org/wiki/Cray.

deeply pipe-lined vector registers that allowed simultaneous parallel operations on functional units.[2]

Whichever computing platform is used in computer models, the finite string retained on all computing machines for the mantissa and exponent means that a floating point number has only an approximate representation in a machine. Furthermore, some real numbers have no machine representation and because modern computing platforms are based on the 64-bit architecture of commodity computers, Sect. 1.9 summarizes their key characteristics in Fortran. The next section explores the consequences of the limitations for numerical accuracy and precision.

1.8 Computer Error and Loss of Significance

From the discussion in Sect. 1.7, it is clear that the accuracy with which numbers are stored in computing machines depends on the word length. If the mantissa of a floating point number is longer than the machine word allows, then the string of coefficients is truncated in a machine- dependent way. Let ε represent the number that is the gap thus created between 1 and $1 + \varepsilon$ which cannot be represented on the machine. When such a case occurs the machine may choose to represent such a number by a floating point number close to it in one of two ways. For example, if the number is $1 + \delta$, where $\delta < \varepsilon$, then its value is chopped to 1 if the excess coefficients, that do not fit into the word length, are simply dropped. Otherwise, $1 + \delta$ is rounded to 1 if $\delta < \frac{1}{2}\varepsilon$ or $1 + \varepsilon$ if $\delta > \frac{1}{2}\varepsilon$. The chopping method ignores all coefficients beyond those stored, but rounding uses the first discarded coefficient to round the last coefficient to be stored. Thus, in this example, chopping would set all floating point numbers between 1 and $1 + \varepsilon$ to the same value of 1 while rounding would do the same for all numbers in the range $1 - \frac{1}{2}\varepsilon$ to $1 + \frac{1}{2}\varepsilon$. Thus, the worst error in machine representation of a number for rounding is $\frac{1}{2}\varepsilon$, while for chopping it is double this, namely ε. The above argument is easily extended to any real floating point number, x, because the machine representation of this number is the perturbation

$$\text{store}(x) = x \times (1 + \varepsilon). \tag{1.20}$$

Thus, in numerical computations, there are several important sources of error due to the way in which numbers are stored in computing machines:

1. error due to the use of store(x) in place of the exact value of the real floating point number, x and,
2. error due to the way in which floating point arithmetic is performed in a computer.

In special circumstances, the compounding effect of these two sources of error leads to propagation and build-up of arithmetic errors in machine computation and much of the design work in numerical algorithms consists of controlling such errors. For a

---

[2] http://pages.cs.wisc.edu/~markhill/restricted/cacm78_cray1.pdf.

1.8 Computer Error and Loss of Significance

detailed examination of this issue, see the book by Higham [4] and Table 2.1 therein for a selection of architectures and their floating point number parameters.

Consider the representation in (1.18) where d_1 is the first significant figure and d_n the nth (or last) significant figure. Clearly, errors in d_1 are more important than errors in d_n if n is much larger than 1, but error propagates from large to small n. Before showing examples of this error propagation, two important measures of error are defined for an approximation y to the exact value x when x, y are real floating point numbers

$$\text{absolute error} = x - y, \tag{1.21}$$

$$\text{relative error} = \frac{x - y}{x}. \tag{1.22}$$

Typical situations where error propagation occurs is in the use of expressions which are not well posed for numerical computation. Examples of such expressions occur for subtraction of nearly equal numbers as in these two examples for small x

$$s(x) = x - sin(x), \tag{1.23}$$

$$e(x) = e^x - e^{-x}, \tag{1.24}$$

or, in this example, for large x

$$r(x) = \sqrt{1 + x} - \sqrt{x}. \tag{1.25}$$

Another example of loss of significance is in a common construct to increment a variable $x = x + h$ when there are a large number of such increments in a loop. The round-off error propagates through d_n from larger to smaller n. Yet another situation similar to these examples occurs in the summation of a series with terms of closely similar magnitude, but alternating in sign. In such cases, to avoid loss of significance due to cancellation between successive terms, positive and negative terms should be summed separately. Other common examples of loss of significance are discussed in subsequent chapters for function and operator approximation where round-off error accumulates and may propagate. For a thorough discussion on precision and errors in computer arithmetic operations, read the ground-breaking expositions in [5,6]. Overton offers an insightful discussion (see Chap. 4 of [6]) of arithmetic operations on legacy and recent microprocessor devices with emphasis on the importance of the IEEE Standard support. The Intel microprocessor floating point unit (FPU) provides a hardware implementation of the IEEE Standard for Binary Floating Point Arithmetic (ANSI/IEEE Std 754-1985) and different compilers offer ways to invoke this standard[3] and they should always be implemented.

EXAMPLES 1.6 to 1.9

---

[3] Always invoke a compiler with the IEEE support option "-mieee-fp" (Intel ifx), or "-Kieee" (Portland Group pgf90).

1.9 Fortran

A successful computer model uses a suite of tools to ensure portability of code across multiple architectures. Contributing scientists write science "modules" as a collection of related subroutines following rules for modules (or functions) in the same class. Then modules in a given class are easily swapped with any other in the same class given suitable interfaces. This procedure for model development presupposes the existence of a suitable human readable machine language with which model algorithms may be coded. Because the legacy algorithms developed by the author over his career are coded in the Fortran language and spanned its evolution as a programming language, a few words about it are appropriate.

Fortran (concatenation of "FORmula TRANslation") was developed as a computer symbolic language in the 1950's and for a summary of the early history see [7]. Since then, it has passed through several standard transitions as defined by the American National Standard Institute (ANSI) and the International Standards Organization (ISO) Programming Language standards. The following is a short summary of the history of the languages with relevant resources as citations:

- 1954 FORTRAN developed for the IBM 704 by a team lead by John Backus[4]
- 1958 FORTRAN II, III
- 1961 FORTRAN IV
- 1966 FORTRAN 66: ANSI BEMA committee
- 1978 FORTRAN 77: ANSI X3.9-1978 [8,9]
- 1997 Fortran 90: ANSI X3.198-1992, ISO/IEC 1539:1991 (E) [10–16][17–19]
- 1995 Fortran 95: ANSI X3J3/96-007, ISO/IEC 1539-1:1996 (Draft) [20,21]
- 2003 Fortran 2003: ISO/IEC 1539-1:2004
- 2010 Fortran 2008: ISO/IEC 1539-1:2010
- 2018 Fortran 2018: ISO/IEC TS 29113:2012, ISO/IEC TS 18508:2015, ISO/IEC/IEEE 60559:2011, IEEE 754-2019 [7]

The past standard in the USA was for both FORTRAN 77 and Fortran 90 and only the latter in Europe. The newer standard was developed over a lengthy period by the ANSI committee X3J3 and the ISO committee SC22/WG5 and is included in the Fortran 90 (and later) standards. These standards include FORTRAN 77 as a subset and code that adheres to the later standard and will compile without modification with current versions. Modern Fortran standards now bring enormous scope and features previously unavailable to Fortran programmers. A full understanding of these features requires a learning effort by programmers wishing to use modern Fortran constructs but examples included with this book do not explore them all.

As the above citations show, many books have been published on the Fortran 90 language and later developments. The Fortran code examples that accompany the algorithms in this book span decades and have their origins in Fortran 66/77. They

---

[4] https://en.wikipedia.org/wiki/John_Backus.

Table 1.1 KIND values for integer and real variables

Valid kind (selected_int_kind): 1=> +/- 10**002
Valid kind (selected_int_kind): 2=> +/- 10**004
Valid kind (selected_int_kind): 4=> +/- 10**009
Valid kind (selected_int_kind): 8=> +/- 10**018
Valid kind (selected_real_kind): 4=> precision=006 range=037
Valid kind (selected_real_kind): 8=> precision=015 range=307

have been updated a little to include features available with commodity processors using the Intel^{TM} Corporation Fortran compiler.[5] Other vendors of Fortran compilers include Portland^{TM},[6] and GNU.[7]

These Fortran vendors have developed compilers to translate Fortran code into machine language for execution on commodity computer platforms. Such compilers invariably adhere to the Fortran standard mentioned above and may include vendor-specific extensions to the standard to implement specific computer architectural features such as the OpenMP standard[8] for thread parallel code algorithms. In the use of Fortran, a point worth noting is that compilers come with many options for compilation ("flags") that control the way in which the compiled code executes on the computer hardware resource. In particular, compiler vendors allow default "optimization" compiler flags to "optimize" execution runtime (i.e., minimize wall clock time). This default choice has a cost in the trade-off in reducing numerical precision and therefore care must be exercised to ensure that the compiler flags chosen offer the best balance of performance time and numerical precision. Several of the author's accompanying Fortran examples include commentary (in the source code) on how numerical precision is enforced to override the default settings.

This chapter closes with the results of a small collection of Fortran examples taken from Chap. 9 of [11] in Tables 1.1, 1.2, 1.3 and 1.4. These investigated a recent 64-bit Intel host processor by using some of the features of Fortran intrinsic functions to explore numerical models. For an explanation of the intrinsic Fortran functions, consult the Fortran Standards documentation [20]. Readers are encouraged to see the output of the Fortran code examples for this chapter on their own host processor.

Table 1.1 used KIND values 1, 2, 4, and 8 for integers and KIND values of 4 and 8, for real numbers to show exponent range and the number of significant figures of precision.

---

[5] http://www.intel.com.
[6] https://www.pgroup.com/, https://developer.nvidia.com/hpc-sdk.
[7] https://gcc.gnu.org/fortran/.
[8] https://www.openmp.org/.

Table 1.2 shows the numeric model for integers with base, number of bits and significant digits, together with the largest value and exponent range.

Table 1.2 Numeric model for integers

Base of model (RADIX): 2
Bits defined in model (BIT_SIZE): 32
Significant digits (DIGITS): 31
Largest value (HUGE): 2147483647
Decimal exponents range (RANGE): 9

Table 1.3 shows the numeric model for real numbers (real *4)

Table 1.3 Numeric model for real numbers

Base of model (RADIX): 2
Significant digits (DIGITS): 24
Minimum exponent (MINEXPONENT): -125
Maximum exponent (MAXEXPONENT): 128
Smallest value (TINY): 0.1175494351E-37
Largest value (HUGE): 0.3402823466E+39
Decimal exponent range(RANGE): 37
Decimal precision (PRECISION): 6
Negligible value (EPSILON): 0.1192092896E-06

Table 1.4 shows the case of a specific real number x with the relative spacing ε on either side.

Table 1.4 Numeric model for real numbers

Value of x: 123.5599976
Exponent part (EXPONENT): 7
Scale (SCALE): 494.2399902
Nearest + infinity (NEAREST): 123.5600052
Nearest − infinity (NEAREST): 123.5599899
Fractional part (FRACTION): 0.9653124809
Absolute spacing (SPACING): 0.7629394531E-05
Relative spacing (RRSPACING): 16195256.00
Exponent form (SET_EXPONENT): 123.5599976

FORTRAN CODE KERRIGEN

1.10 Exercises

(a) Perform a conversion to both decimal and binary for the hexadecimal numbers

   a. 1F.C
   b. 11.1

(b) Consider a computing machine that stores numbers in binary form in normalized representation with six bit words such that 1 bit is for the sign (0 positive and 1 negative), 3 bits for the exponent (sign and magnitude), and 2 bits for the mantissa. What is

   (a) the binary string and decimal value of the first four numbers of smallest magnitude?
   (b) the gap (in decimal value) between each of the four numbers computed in (a)?
   (c) the binary string and decimal value of the largest positive number that can be stored in the machine?

1.11 Programming Problems

1. Write a program in the coding language of your choice, Fortran, C, or Julia for any (or all) of the algorithms in EXAMPLES 1.2 to 1.5.
2. Compile the FORTRAN CODE KERRIGEN on your platform and execute it. Compare the output you see against the results of Tables 1.1 to 1.4.
3. If you have access to an HPC resource, repeat problem 2 with that platform.

References

1. Stone JV (2015) Information theory a tutorial introduction. Sebtel Press
2. Delic G, Cash GG (2000) The permanent of 0,1 matrices and kallman's algorithm. Comput Phys Commun 124:315–329
3. Copeland BJ (ed) (2006) Colossus: the secrets of Bletchly Park's code-breaking computers. Oxford University Press, Oxford, UK
4. Higham NJ (1996) Accuracy and stability of numerical algorithms. Society for Industrial and Applied Mathematics, Philadelphia, PA
5. Wilkinson JH (1963) Rounding errors in algebraic processes. Prentice-Hall Inc, Englewood Cliffs, NJ

6. Metcalf M, Reid J, Cohen M (2018) Numerical computing with IEEE floating point arithmetic. Society for Industrial and Applied Mathematics, Philadelphia, PA
7. Metcalf M, Reid J, Cohen M (2018) Modern Fortran explained. Oxford University Press, Oxford, UK
8. Metcalf M (1985) Effective Fortran 77. Oxford University Press, New York, NY
9. Borse GJ (1985) Fortran 77 and numerical methods for engineers. PWS-Kent Publishing, Boston, MA
10. Adams Jeanne C, Brainerd Walter S, Martin Jeanne T, Smith Brian T, Wagener Jerrold L (1979) Fortran 90 handbook complete ANSI/ISO reference. Intertext Publications/McGraw-Hill Book Company, New York, NY
11. Kerrigan JF (1993) Migrating to Fortran 90. O'Reilly and Associates, Sebastopol, CA
12. Chivers I, Sleightholme J (1995) Introducing Fortran 90. Springer-Verlag, London, UK
13. Gehrke W (1996) Fortran 90 language guide. Springer-Verlag, London, UK
14. Hahn Brian D (1994) Fortran 90 for scientists and engineers. Edward Arnold/Hodder Headline Group, London, UK
15. Mayo WE, Cwiakala M (1996) Theory and problems of programming with Fortran 90. McGraw Hill, New York, NY
16. Metcalf M, Reid J (1990) Fortran 90 explained. Oxford University Press, Oxford, UK
17. Redwine C (1995) Upgrading to Fortran 90. Springer-Verlag, New York, NY
18. Brainerd WS, Goldberg CH, Adams JC (1996) Programmer's guide to Fortran 90, 3rd edn. Springer-Verlag, New York, NY
19. Brooks DR (1997) Problem solving with Fortran 90. Springer-Verlag, New York, NY
20. Adams JC, Brainerd WS, Martin JT, Smith BT, Wagener JL (1997) Fortran 95 handbook complete ANSI/ISO reference. The MIT Press, Cambridge, MA
21. Metcalf M, Reid J (1996) Fortran 90/95 explained. Oxford University Press, Oxford, UK

Function Approximation and Error

2.1 Function Approximation by Recurrence

Recurrence is frequently an efficient means of producing approximations to function values if it is stable (i.e., numerical error does not grow). A simple example is that for evaluation of the series expansion

$$x - \sin(x) = \frac{x^3}{3!} - \frac{x^5}{5!} + \frac{x^7}{7!} - \frac{x^9}{9!} + \cdots, \tag{2.1}$$

where a two-term recurrence for successive terms is

$$t_{n+1} = -\frac{x^2}{(2n+2)(2n+3)} t_n, \quad n = 1, 2, \ldots, \tag{2.2}$$

with an initial value as the first term of (2.1)

$$t_1 = \frac{x^3}{3!}, \tag{2.3}$$

and the sequence of partial sums of the series is $s_{n+1} = s_n + t_n$, with $s_1 = 0$. However, recurrence is not always accurate or stable and each new case should be studied separately.

A simple example of a three-term recurrence is that for the spherical Bessel function $j_\ell(r)$ of integer order ℓ and real argument r, discussed in Sect. 2.6, that satisfies

$$j_{\ell-1}(r) = \frac{2\ell+1}{r} j_\ell(r) - j_{\ell+1}(r). \tag{2.4}$$

Supplementary Information The online version contains supplementary material available at https://doi.org/10.1007/978-3-031-90178-2_2

© The Author(s), under exclusive license to Springer Nature Switzerland AG 2026
G. Delic, *Guide to Numerical Algorithm Design and Development*,
Texts in Computer Science, https://doi.org/10.1007/978-3-031-90178-2_2

The recurrence is started at some order, L, greater than the highest order needed, with a starting value given by either of these asymptotic expressions

$$j_\ell(r) \approx \sqrt{\frac{e}{2}} \frac{(er)^\ell}{(2\ell+1)^{\ell+1}},$$

$$j_\ell(r) \approx \sqrt{\frac{\pi}{\ell}} \frac{1}{\ell!} \left(\frac{r}{2}\right)^\ell e^{-r^2/2\ell}, \ r \ll \ell, \quad (2.5)$$

where e is the base of the natural logarithm, a mathematical constant and the second form of (2.5) is from p344 of [1], see also [2]. A normalization constant is obtained from the result

$$\sum_{\ell=0}^{\infty} (2\ell+1) j_\ell^2(r) = 1, \quad (2.6)$$

where the upper limit in the summation is set to L. Accuracy obtained by this algorithm can then be compared to the exact values for the two lowest orders evaluated at argument r

$$j_0(r) = \frac{\sin(r)}{r}, \quad (2.7)$$

$$j_1(r) = \frac{1}{r}\left[\frac{\sin(r)}{r} - \cos(r)\right]. \quad (2.8)$$

The recurrence in order of (2.4) can become problematic for at least three reasons

- it is stable only when ℓ is running from higher to lower order,
- it is sensitive to the starting value of ℓ and
- accuracy is acceptable at smaller values of r, but less so at larger values.

On the other hand, similar three-term recurrences for polynomials are known to be stable for increasing order and fixed argument (see Sect. 2.4). An alternative (but compute-intensive) method of obtaining values of $j_\ell(r)$ is application of the series expansion

$$j_\ell(r) = \frac{\sqrt{\pi}}{2^{\ell+1}} r^\ell \sum_{k=0}^{\infty} c(\ell + \tfrac{1}{2}, k) r^{2k}, \quad (2.9)$$

where the coefficient is

$$c(\ell + \tfrac{1}{2}, k) = \frac{(-1)^k}{4^k k! \Gamma(\ell + k + \tfrac{3}{2})}, \quad (2.10)$$

and $\Gamma(n + \tfrac{1}{2})$ is the Euler Gamma function for integer n, with properties

$$\begin{aligned} \Gamma(n+\tfrac{3}{2}) &= \left(n + \tfrac{1}{2}\right) \Gamma(n + \tfrac{1}{2}), \\ \Gamma(n+\tfrac{1}{2}) &= \tfrac{1}{2}\tfrac{1}{2} + 1\tfrac{1}{2} + 2 \cdots \tfrac{1}{2} + (n-1)\Gamma(\tfrac{1}{2}), \\ \Gamma(\tfrac{1}{2}) &= \sqrt{\pi}. \end{aligned} \quad (2.11)$$

A two-term recurrence for successive coefficients in (2.9) is

$$c(\ell + \tfrac{1}{2}, k+1) = \frac{-1}{(2k+2)(2\ell + 2k + 3)} c(\ell + \tfrac{1}{2}, k). \qquad (2.12)$$

The case of the spherical Bessel function is discussed in Sect. 2.6 with a Fortran code application.

EXAMPLES 1.8 and 1.9

2.2 Approximation with a Set of Basis Functions

In the previous section, a series expansion was introduced to approximate a function where it is assumed that the function itself is smooth (i.e., has continuous derivatives). The nature of the approximation depends on the choice of a set of basis elements $\{\psi_K\}_1^\infty$ for a discrete index $K = 1, 2, 3, \ldots$. The basis functions (or elements) are likewise assumed be to smooth functions of the argument. In Chap. 4 a detailed exposition covers concepts regarding the use of basis functions, the spaces they span and means of measuring the quality of the approximations they provide. Along the way several types of basis functions are introduced and their utility discussed in the context of applications. For now, consider a function of one variable $f(x)$, on the interval $x \in [-1, +1]$, replaced with a linear series of basis functions as follows:

$$f(x) = \sum_{K=1}^\infty c_K \psi_K(x). \qquad (2.13)$$

The interval need not be restricted to $x \in [-1, +1]$, because mappings may be used to change the variable, as in the simple linear map

$$x \in [a, b],$$
$$t \in [-1, +1],$$
$$x = \frac{1}{2}(b+a) + \frac{1}{2}(b-a)t,$$
$$t = \frac{2x - (b+a)}{(b-a)}. \qquad (2.14)$$

However, for simplicity and for the sake of the following discussion, we assume the normalized interval $x \in [-1, +1]$.

Throughout this book, the effect of truncating the series at a finite value of $K = N$ is explored, with reference to the resulting error on the interval of the argument's range. An associated issue is the question: how does the choice of basis set affect the error of the truncated series? In essence, a different choice of basis results in different errors for the same number of terms. Also, a pertinent question is: which choice of basis gives the "best approximation"? This can only be completely answered when the meaning of "best approximation" is defined and this requires a precise measure of assessing the error introduced in truncation of the series. This is the subject of Chap. 4 where different methods of precisely measuring the error in the truncated series are defined. For the present discussion, the truncation error is measured by the error curve defined over the whole interval $x \in [-1, +1]$,

$$E_N(x) = f(x) - \sum_{K=1}^{N} c_K \psi_K(x). \quad (2.15)$$

The error curve is different for each choice of N but, it is expected that, at each x, $E_N(x) \to 0$ as $N \to \infty$. Often there is interest only in the worst possible error when $|E_N(x)|$ reaches a maximum and therefore it suffices to consider only that value of x for which this occurs. From this estimate, a bound on the error is obtained as

$$\max_{x \in [-1,+1]} |E_N(x)|. \quad (2.16)$$

Throughout this book, methods of estimating the "maximum error" is a recurring theme. Clearly, the choice of basis that minimizes the "maximum error" would be considered the most successful and therefore deserves more study with a view to applications. In the remainder of this chapter, a few possible choices of basis set are described and examples of applications are discussed to elucidate some of the issues introduced above.

EXAMPLES 2.1 and 2.2

2.3 Monomial Basis Functions

Any complete set of functions may be chosen as a basis to approximate a general function, but they must satisfy basic criteria (to be developed as this discussion progresses). A simple basis is the set of monomials. These are integer powers of the argument and constitute a basis set as follows:

$$\psi_K(x) = x^{K-1}, \ K = 1, 2, 3, \ldots, \quad (2.17)$$

where $x \in [-1, +1]$ is a finite interval. Then, with a change in variable for the index, $k = K - 1$, the index ranges $k = 0, 1, 2, \ldots$. Care should be taken to determine whether the index begins at zero, one, or some other value. The choice of monomials, $\{\psi_k\}_0^\infty$, as basis elements leads to a simple power series for any continuous smooth function $f(x)$ defined on the same interval as x

$$f(x) = \sum_{k=0}^{\infty} c_k x^k, \quad (2.18)$$

2.3 Monomial Basis Functions

where the coefficients, c_k, are different for different functions $f(x)$. One method of obtaining the unknown coefficients c_k is from the Taylor series, which is an expansion in powers of the variable x, minus a constant value x_0. This can be rearranged as a monomial series, such as in (2.18). However, the usual definition of a Taylor series [3] is

$$f(x) = f(x_0) + f^{(1)}(x_0)[x - x_0] + \frac{1}{2!}f^{(2)}(x_0)[x - x_0]^2 + \frac{1}{3!}f^{(3)}(x_0)[x - x_0]^3 + \cdots \quad (2.19)$$

$$+ \frac{1}{k!}f^{(k)}(x_0)[x - x_0]^k + \cdots,$$

where there are an infinite number of terms. Obviously, the function must be infinitely differentiable (or smooth) on a bounded interval, $x \in [a, b]$, to apply this method of obtaining expressions for the coefficients, c_k, in (2.18). In (2.19), the notation $f^{(k)}(x_0)$ indicates that the kth derivative of the function is evaluated at x_0 with $f^{(0)} \equiv f$. The notation $k!$ is the factorial function defined as $k! = 1 \times 2 \times 3 \times \cdots \times k$, with $0! = 1! = 1$. A concise representation of the result in (2.19) is

$$f(x) = \sum_{k=0}^{\infty} \frac{1}{k!} f^{(k)}(x_0)[x - x_0]^k, \quad (2.20)$$

and, when $x_0 = 0$, this last result becomes the MacLaurin series

$$f(x) = \sum_{k=0}^{\infty} \frac{1}{k!} f^{(k)}(0) x^k. \quad (2.21)$$

Specific examples of Taylor series, with $x_0 = 0$, for some standard functions, are the following for $-\infty < x < +\infty$,

$$e^x = 1 + x + \frac{1}{2!}x^2 + \frac{1}{3!}x^3 + \frac{1}{4!}x^4 + \cdots + \frac{1}{k!}x^k + \cdots, \quad (2.22)$$

$$sin(x) = x - \frac{1}{3!}x^3 + \frac{1}{5!}x^5 - \frac{1}{7!}x^7 + \cdots + \frac{(-1)^{k+1}}{(2k-1)!}x^{2k-1} + \cdots, \quad (2.23)$$

$$cos(x) = 1 - \frac{1}{2!}x^2 + \frac{1}{4!}x^4 - \frac{1}{6!}x^6 + \cdots + \frac{(-1)^k}{(2k)!}x^{2k} + \cdots, \quad (2.24)$$

while for $-1 < x < +1$

$$\frac{1}{1-x} = 1 + x + x^2 + x^4 + \cdots + x^k + \cdots, \quad (2.25)$$

and

$$\frac{1}{(1-x)^n} = 1 + nx + \binom{n+1}{2}x^2 + \binom{n+2}{3}x^3 + \cdots + \binom{k+n-1}{n-1}x^k + \cdots \quad (2.26)$$

where the binomial coefficients are defined by (3.67), also for $|x-1| < 1, k > 0$

$$ln(x) = (x-1) - \frac{1}{2}(x-1)^2 + \frac{1}{3}(x-1)^3 - \frac{1}{4}(x-1)^4 + \cdots + \frac{(-1)^{k+1}}{k}(x-1)^k + \cdots. \tag{2.27}$$

Note that e^x has both even and odd powers in argument x, while $sin(x)$ and $cos(x)$, have series of only odd and even powers, respectively. This is because $sin(x)$ is an odd function with respect to the origin and $f(x) = -f(-x)$, while $cos(x)$ is even with $f(x) = f(-x)$. An odd function cannot be expanded in even powers of the argument and vice versa.

All the series given in (2.22)–(2.27) converge to the exact value of the corresponding function for the respective intervals in argument. However, if the series on the right-hand side of (2.19) is truncated at the nth derivative, then the resulting function is the Taylor polynomial of order n. In this case, the error given by (2.15), (2.16) for the truncated series is known to be

$$E_{n+1} = \frac{1}{(n+1)!} f^{(n+1)}(\xi)[x - x_0]^{n+1}, \tag{2.28}$$

where the derivative term is evaluated at some point ξ which is in the interval $x_0 < \xi < x$. Because the interval size depends on the choice of x, then ξ is a function of x, and is written $\xi(x)$. Examples of explicit expressions of the error for the series of (2.22)–(2.27) are shown in Table 2.1. In (2.28) the value of the derivative at ξ may be thought of as an average value in the interval $[x_0, x]$, as is demonstrated by the special case of $n = 0$ when

$$E_1 = f^{(1)}(\xi)[x - x_0], \tag{2.29}$$

and the Taylor series is

$$f(x) \approx f(x_0) + E_1. \tag{2.30}$$

From the last two equations, it follows that

$$f^{(1)}(\xi) = \frac{f(x) - f(x_0)}{x - x_0}, \tag{2.31}$$

which is seen to be the average value of the first derivative on the interval $[x_0, x]$. Substituting (2.29) into (2.30) gives the Mean Value Theorem for this special case. This theorem states that a function which is continuous on an interval $[x_0, x]$ also possesses a derivative on that interval.

A special case of the Taylor series, (2.20), occurs if $x - x_0 = h$, a constant, then with the substitution $z = x_0$, it follows

$$f(z+h) = \sum_{k=0}^{\infty} \frac{1}{k!} f^{(k)}(z) h^k + E_{n+1} \tag{2.32}$$

2.3 Monomial Basis Functions

Table 2.1 Function and error term

e^x	$E_{n+1} = \frac{x^{n+1}}{(n+1)!} e^{\xi}$
$\sin(x)$	$E_{2n+3} = (-1)^{n+1} \frac{x^{2n+3}}{(2n+3)!} \cos(\xi)$
$\cos(x)$	$E_{2n+2} = (-1)^{n+1} \frac{x^{2n+2}}{(2n+2)!} \cos(\xi)$
$\frac{1}{1-x}$	$E_{n+1} = \frac{x^{n+1}}{1-\xi}, \xi \neq 1$
$\ln(1+x)$	$E_{n+1} = (-1)^n \frac{x^{n+1}}{(n+1)} \frac{1}{(1+\xi)^{n+1}}$

with the error term

$$E_{n+1} = \frac{f^{(n+1)}(\xi)}{(n+1)!} h^{n+1}, \; z < \xi < z+h. \tag{2.33}$$

The result of (2.32) is a power series in h and furthermore, (2.33) shows that the behavior of the error is $E_{n+1} \propto h^{n+1}$, with an estimate on the bound given by $|E_{n+1}| \leq C|h|^{n+1}$ for a finite constant, C, such that

$$\frac{|f^{(n+1)}(\xi)|}{(n+1)!} \leq C. \tag{2.34}$$

The special result of (2.33) is summarized by saying that $E_{n+1} \to 0$ as the $n+1$ power of h. When $h \to 0$ this is denoted as

$$E_{n+1} \approx \mathcal{O}(h^{n+1}) \tag{2.35}$$

where $\mathcal{O}$ means "of the order of".

Consider an example in the quadratic approximation to $f(z+h)$

$$f(z+h) = f(z) + f^{(1)}(z)h + \frac{1}{2!} f^{(2)}(z)h^2 + \frac{1}{3!} f^{(3)}(\xi)h^3. \tag{2.36}$$

The error is $\mathcal{O}(h^3)$ which means that accuracy in the value of $f(z+h)$, computed from the left-hand side of (2.36), has cubic convergence as $h \to 0$.

While monomials are useful in introducing the idea of approximation of functions, it is much more practical, from the point of view of accuracy and computational efficiency, to use polynomials. These are defined in the next section.

EXAMPLES 2.3–2.6

Table 2.2 Two examples of polynomials

Legendre	Chebyshev
$P_0(x) = 1$	$T_0(x) = 1$
$P_1(x) = x$	$T_1(x) = x$
$P_2(x) = \frac{1}{2}(3x^2 - 1)$	$T_2(x) = 2x^2 - 1$
$P_3(x) = \frac{1}{2}(5x^3 - 3x)$	$T_3(x) = 4x^3 - 3x$
$P_4(x) = \frac{1}{8}(35x^4 - 30x^2 + 3)$	$T_4(x) = 8x^4 - 8x^2 + 1$
$P_5(x) = \frac{1}{8}(63x^5 - 70x^3 + 15x)$	$T_5(x) = 16x^5 - 20x^3 + 5x$

2.4 Polynomial Basis Functions

A theorem of Weierstrass[1] states that, for a continuous function on a finite interval of the argument, a polynomial of sufficiently high order exists, such that it approaches the function value in that range [4]. How this approach is measured is discussed in more detail in Chap. 4. For the moment, in this section, polynomial basis sets are introduced that are used later in the approximation of functions. These are the Chebyshev [4,5] and Legendre polynomials [6,7] defined on the interval $x \in [-1, +1]$. First a definition: a power series of monomials, in argument x, that terminates at the $(n + 1)$th term, is called a polynomials $p_n(x)$ of order n

$$p_n(x) = \sum_{k=0}^{n} c_k x^k. \qquad (2.37)$$

Note that each term of (2.37), $c_k x^k$, is a monomial. Several types of the classical polynomials have been studied and each type corresponds to a different choice of the set of coefficients $\{c_k\}_0^n$. In this book, emphasis is placed on two special cases: the Legendre polynomials, which have their origin in classical potential theory (Chap. 24 of [8]) and Chebyshev polynomials, named after their discoverer[2] (sometimes spelled as Tchebychef), whose complete works are available in French [9]. The Chebyshev polynomials are important because of their remarkable properties in numerical approximation [5,10]. Both of these polynomials are defined on the interval $x \in [-1, +1]$ and the first few of each type are listed in Table 2.2. A graphical representation may be found in the book by Spanier and Oldham [11] and in some of the examples.

Unlike monomials, these polynomials oscillate between negative and positive values on the interval $x \in [-1, +1]$ and the roots of a polynomial correspond to those values of x for which $p_n(x) = 0$. The roots are distinct for different order n,

---

[1] Karl Wilhelm Theodor Weierstrass, German mathematician 1815–1897.
[2] Russian mathematician P.I. Chebyshev (1821–1894).

2.4 Polynomial Basis Functions

Table 2.3 Special values of polynomials

Legendre	Chebyshev
$P_k(1) = 1$	$T_k(1) = 1$
$P_k(-1) = (-1)^k$	$T_k(-1) = (-1)^k$
$P_{2j}(0) = (-1)^j \prod_{i=0}^{j-1} \frac{(2i+1)}{(2i+2)}$	$T_{2j}(0) = (-1)^j$
$P_{2j+1}(0) = 0$	$T_{2j+1}(0) = 0$

except at the origin (see below). For this reason, values of x for which this occurs, are sometimes called zeros. The zeros of the Chebyshev polynomial $T_n(x)$ are located at

$$x_{n-j} = \cos\left[\frac{2j+1}{2n} \times \pi\right], \quad j = 0, 1, \ldots, n-1, \tag{2.38}$$

where the index orders the roots on the interval $x \in [-1, +1]$ according to the ranking

$$-1 < x_1 < x_2 < \cdots < x_{n-1} < x_n < +1. \tag{2.39}$$

However, the zeros of $P_n(x)$ are not known analytically and must be found by numerical algorithms such as those discussed in Chap. 6. The zeros of Legendre and Chebyshev polynomials are all real numbers and distinct. They have the interlacing property, which means that, with the exception of the roots closest to $x = \pm 1$, the zeros of the polynomial of order $n+1$ lie between the zeros of the polynomial of order n.

The polynomials are symmetric with respect to the origin $x = 0$, whence

$$P_k(-x) = (-1)^k P_k(x), \tag{2.40}$$

$$T_k(-x) = (-1)^k T_k(x), \tag{2.41}$$

and a few special values at $x = \pm 1$, or $x = 0$, are given in Table 2.3, with the definition

$$\prod_{i=0}^{k} \frac{(2i+1)}{(2i+2)} = \frac{(1)}{(2)} \frac{(3)}{(4)} \cdots \frac{(2k+1)}{(2k+2)}. \tag{2.42}$$

The polynomials satisfy a three-term recurrence relation that may be used to generate all higher order polynomials, at a fixed value of $x \in [-1, +1]$, given only values of the two lowest order polynomials. For the Legendre case, using the values of $P_0(x) = 1$, $P_1(x) = x$, the recurrence for $n = 1, 2, 3, \ldots$, is

$$(n+1)P_{n+1}(x) = (2n+1)x P_n(x) - n P_{n-1}(x). \tag{2.43}$$

Once all the polynomial values up to order n are available, the first derivative of the polynomial, with respect to argument, may also be generated from a three-term recurrence. Thus, for the Legendre polynomials, with $n = 1, 2, 3, \ldots$, the recurrence is

$$(1-x^2)\frac{dP_n(x)}{dx} = n\left[P_{n-1}(x) - xP_n(x)\right]. \tag{2.44}$$

Similarly, for Chebyshev polynomials, with $T_0(x) = 1$, $T_1(x) = x$, the recurrence for $n = 1, 2, 3, \ldots$, is

$$T_{n+1}(x) = 2xT_n(x) - T_{n-1}(x). \tag{2.45}$$

Also, the first derivative of the Chebyshev polynomial, with respect to argument, is generated from the three-term recurrence

$$(1-x^2)\frac{dT_n(x)}{dx} = n\left[T_{n-1}(x) - xT_n(x)\right]. \tag{2.46}$$

The polynomials $P_n(x)$ and $T_n(x)$ are said to be orthogonal because they satisfy orthogonality conditions of the following type: for the Legendre case

$$\int_{-1}^{+1} P_m(x)P_n(x)dx = 0, \, m \neq n,$$

$$= \frac{2}{2n+1}, \, m = n, \tag{2.47}$$

and for the Chebyshev case

$$\int_{-1}^{+1} T_m(x)T_n(x)\frac{dx}{\sqrt{1-x^2}} = 0, \, m \neq n,$$

$$= \frac{\pi}{2}, \, m = n \neq 0,$$

$$= \pi, \, m = n = 0. \tag{2.48}$$

Polynomials provide an important class of basis functions and they have a fundamental role in numerical approximation [5, 10, 12]. This is exemplified by the special choice of basis set in (2.13) as one of these two

$$\psi_K(x) = P_{K-1}(x), \tag{2.49}$$
$$\psi_K(x) = T_{K-1}(x). \tag{2.50}$$

The corresponding series expansion, analogous to (2.13), for the Legendre polynomials is

$$f(x) = \sum_{k=0}^{\infty} (k + \frac{1}{2}) g_k P_k(x), \tag{2.51}$$

2.4 Polynomial Basis Functions

where the coefficients of (2.51) are defined by

$$g_k = \int_{-1}^{+1} f(x) P_k(x) dx, \qquad (2.52)$$

and for the Chebyshev polynomials the series expansion is

$$f(x) = \sum_{k=0}^{\infty\prime} h_k T_k(x), \qquad (2.53)$$

where the prime indicates that the first term of the series includes a factor of $\frac{1}{2}$ and reads as $\frac{1}{2} h_0 T_0(x)$. The Chebyshev coefficients of (2.53) are defined by

$$h_k = \frac{2}{\pi} \int_{-1}^{+1} f(x) T_k(x) \frac{dx}{\sqrt{1-x^2}}. \qquad (2.54)$$

The expressions for the coefficients follow on applying the corresponding orthogonality condition of (2.47), (2.48). This result follows, on multiplying both sides of the respective series expansions (2.51), (2.53) by the corresponding polynomial of any order and choosing the corresponding weight function as one of the following

$$\varphi(x) = 1, \qquad (2.55)$$

$$\varphi(x) = \frac{1}{\sqrt{1-x^2}}. \qquad (2.56)$$

This technique is demonstrated here, in the Chebyshev case, with the series expansion of (2.53)

$$\int_{-1}^{+1} f(x) T_m(x) \frac{dx}{\sqrt{1-x^2}} = \sum_{k=0}^{\infty\prime} h_k \int_{-1}^{+1} T_k(x) T_m(x) \frac{dx}{\sqrt{1-x^2}}, \qquad (2.57)$$

where only the $k = m$ term survives on the right-hand side because of the orthogonality property of (2.48). Application of the definitions in (2.52), (2.54) for g_k, h_k, respectively, shows that monomials on $x \in [-1, +1]$ have an expansion in polynomials of either type and some examples are shown in Table 2.4.

If the coefficients of a monomial series are known, as in (2.22) to (2.26), then a simple result of Rivlin [5, 10] in Exercise 1.2.6, may be applied to obtain the Chebyshev series expansion coefficients. This is shown in EXAMPLE 2.33 when this method is applied for $exp(x)$ in (2.22).

An alternative approach computes the Chebyshev series coefficients in the expansion for $exp(zx)$. Luke [13] on page 43, offers a Chebyshev series expansion on the interval $x \in [-1, +1]$, written here in the form

$$e^{zx} = \sum_{k=0}^{\infty} \epsilon_k I_k(z) T_k(x), \quad \epsilon_0 = 1, \epsilon_k = 2, k > 0, \qquad (2.58)$$

Table 2.4 Polynomial expansions

Legendre	Chebyshev
$x^0 = P_0(0)$	$x^0 = T_0(0)$
$x^1 = P_1(x)$	$x^1 = T_1(x)$
$x^2 = \frac{1}{3}[P_0(x) + 2P_2(x)]$	$x^2 = \frac{1}{2}[T_0(x) + T_2(x)]$
$x^3 = \frac{1}{5}[3P_1(x) + 2P_3(x)]$	$x^3 = \frac{1}{4}[3T_1(x) + T_3(x)]$
$x^4 = \frac{1}{35}[7P_0(x) + 20P_2(x) + 8P_4(x)]$	$x^4 = \frac{1}{8}[3T_0(x) + 4T_2(x) + T_4(x)]$
$x^5 = \frac{1}{63}[27P_1(x) + 28P_3(x) + 8P_5(x)]$	$x^5 = \frac{1}{16}[10T_1(x) + 5T_3(x) + T_5(x)]$

where $I_k(z)$ is the Modified Bessel Function of the first kind of integer order defined in [6], where the explicit form is given as a series in 9.6.10 (page 375)

$$I_k(z) = \left(\frac{z}{2}\right)^k \sum_{j=0}^{\infty} \frac{\left(\frac{z^2}{4}\right)^j}{j!\Gamma(k+j+1)}. \tag{2.59}$$

Here the Gamma function that appeared in (2.11) has integer argument, with $\Gamma(n+j+1) = (n+j)!$. When $z = 1$, the series in (2.59) simplifies to the form

$$I_k(1) = \frac{1}{2^k} \sum_{j=0}^{\infty} c(k, j), \tag{2.60}$$

where

$$c(k, j) = \frac{1}{4^j} \frac{1}{j!\Gamma(k+j+1)}. \tag{2.61}$$

With a starting value of $c(k, 0) = \frac{1}{k!}$, the coefficient satisfies a simple (upward) recurrence

$$c(k, j+1) = \frac{c(k, j)}{4(j+1)(k+j+1)}. \tag{2.62}$$

The series (2.60) converges rapidly as is demonstrated in EXAMPLE 2.34.

Properties specific to Chebyshev polynomials make for easy evaluation of a truncated series such as

$$f_N(x) = \sum_{k=0}^{N\prime} h_k T_k(x), \tag{2.63}$$

which may be evaluated by recurrence for fixed x using $A_{N+2} = A_{N+1} = 0$ and computing a sequence for $n = N, N-1, \ldots, 0$, in

$$A_n = 2x A_{n+1} - A_{n+2} + h_n. \tag{2.64}$$

Then the value of the series in (2.63) is given by

$$f_N(x) = A_0 - xA_1 - \frac{1}{2}h_0. \tag{2.65}$$

If the Chebyshev series consists of only even polynomials as in

$$f_N(x) = \sum_{k=0}^{N\prime} h_k T_{2k}(x), \tag{2.66}$$

a modified recurrence is applied

$$A_n = 2(2x^2 - 1)A_{n+1} - A_{n+2} + h_n, \tag{2.67}$$

and the value of the series in (2.66) is given by

$$f_N(x) = A_0 - (2x^2 - 1)A_1 - \frac{1}{2}h_0. \tag{2.68}$$

For a Chebyshev series consisting of only odd polynomials, as in

$$f_N(x) = \sum_{k=0}^{N} h_k T_{2k+1}(x), \tag{2.69}$$

the same recurrence in (2.67) is applied, but the numerical series value is obtained from

$$f_N(x) = x(A_0 - A_1). \tag{2.70}$$

It should be noted that the meaning of N is different in the three series of (2.63), (2.66), (2.69), as is the subscript of the Chebyshev coefficient h_k. In fact the value of the series (2.63) could be obtained by evaluating even and odd terms separately in (2.66), (2.69).

The results of (2.63)–(2.70) follow on application of the Chebyshev polynomial recurrence relation (2.45) to the respective series of (2.63), (2.66), (2.69). Note that in the series (2.63), (2.66), the coefficient of the first term h_0 does not incorporate the factor $\frac{1}{2}$ denoted by the prime in the summation, as is done by some authors. Special care should be exercised with respect to this factor in applications. For additional detail see Chap. 8 and Sect. 8.5 of [14], volume I, or Chap. 3 of [5,10].

EXAMPLES 2.7–2.13 and 2.36

2.5 Trigonometric Basis Functions

Trigonometric basis functions, or Fourier series [15], are usually applied if the function to be approximated is known to be periodic with $f(x + T) = f(x)$, for a fixed period $T > 0$. In this case, an appropriate choice of basis functions are trigonometric functions such as sine or cosine. If the function to be approximated is either even, $f(-x) = f(x)$, or odd, $f(-x) = -f(x)$, with respect to the origin, $x = 0$, for a symmetric interval, then either a cosine or sine series, respectively, should be applied. This is because an even function can only have even powers of the argument in a power series, while only odd powers occur for an odd function as is the case, for example, with $cos(x)$ and $sin(x)$.

An arbitrary function $f(x)$ may be expanded in Fourier series of argument $x \in [-1, +1]$

$$f(x) = \sum_{k=0}^{\infty\prime} c_k cos(k\pi x) + \sum_{k=1}^{\infty} s_k sin(k\pi x), \tag{2.71}$$

with the basis functions

$$cos(0\pi x), sin(1\pi x), cos(1\pi x), \ldots, sin(k\pi x), cos(k\pi x), \ldots, \tag{2.72}$$

where, as before, the prime indicates that the first term has included a factor of $\frac{1}{2}$ and is written as $\frac{1}{2}c_0 cos(0\pi x)$. On introduction of the variable $\theta = \pi x$, definitions of the coefficients are given by

$$c_k = \frac{1}{\pi} \int_{-\pi}^{+\pi} f(\theta) cos(k\theta) d\theta, \ k = 0, 1, 2, 3, \ldots, \tag{2.73}$$

$$s_k = \frac{1}{\pi} \int_{-\pi}^{+\pi} f(\theta) sin(k\theta) d\theta, \ k = 1, 2, 3, \ldots. \tag{2.74}$$

The expressions for the coefficients follow on multiplying the series expansion by the respective trigonometric functions and applying the orthogonality relations

$$\frac{1}{\pi} \int_{-\pi}^{+\pi} cos(n\theta) cos(m\theta) d\theta = \delta_{nm}, \tag{2.75}$$

$$\frac{1}{\pi} \int_{-\pi}^{+\pi} sin(n\theta) sin(m\theta) d\theta = \delta_{nm}, \tag{2.76}$$

$$\frac{1}{\pi} \int_{-\pi}^{+\pi} cos(n\theta) sin(m\theta) d\theta = 0, \ all \ n, m, \tag{2.77}$$

where the Kronecker[3] delta function is defined by

$$\delta_{nm} = 1, m = n,$$
$$= 0, m \neq n. \tag{2.78}$$

---

[3] Leopold Kronecker, German mathematician 1823–1891.

2.5 Trigonometric Basis Functions

Although they are rarely used in numerical applications, the trigonometric functions also satisfy recurrence relations,

$$cos[(n+1)\pi x] = cos(n\pi x)cos(\pi x) - sin(n\pi x)sin(\pi x), \quad (2.79)$$
$$sin[(n+1)\pi x] = sin(n\pi x)cos(\pi x) + cos(n\pi x)sin(\pi x), \quad (2.80)$$

and are applied with the starting values

$$cos(1\pi x), \quad (2.81)$$
$$sin(1\pi x). \quad (2.82)$$

A truncated cosine series

$$C_N(\theta) = \sum_{k=0}^{N'} c_k cos(k\theta), \quad (2.83)$$

may be evaluated numerically, for fixed θ, by recurrence using $A_{N+2} = A_{N+1} = 0$, for $n = N, N-1, \ldots, 0$, in

$$A_n = 2cos(\theta)A_{n+1} - A_{n+2} + c_n, \quad (2.84)$$

and then the series value is

$$C_N(\theta) = A_0 - cos(\theta)A_1 - \frac{1}{2}c_0. \quad (2.85)$$

Similarly, a truncated sine series

$$S_N(\theta) = \sum_{k=1}^{N} s_k sin(k\theta), \quad (2.86)$$

may be evaluated by the same recurrence in (2.84) with s_n in place of c_n but terminating at $n = 1$, when the series value is given by

$$S_N(\theta) = A_1 sin(\theta). \quad (2.87)$$

The methods described in this (and the previous) section for evaluation of truncated series are based on the algorithm developed by Clenshaw [16,17] for Chebyshev series. It was subsequently also applied in the solution of differential equations [18] by Clenshaw and in the numerical integration of functions [19] by Clenshaw and Curtis. The algorithm is more generally applicable when the basis functions have orthogonality properties and satisfy recurrence relations. For additional detail see Chap. 3 of Fox and Parker [4] and Sect. 5.5 in [20]. A diverse collection of Chebyshev expansions is given by Luke [13,21] complete with numerical tables of coefficients and Fortran code for generating them.

Fourier series have a long history since their discovery and application by Joseph Fourier[4] in the early 1800's [22] and suitable resources about theory and applications abound for both continuous and discrete functions [15,23–29]. A collection of expansion formulas for a range of continuous function is given in [30].

EXAMPLES 2.14–2.18

2.6 Application to Bessel Functions

As a demonstration of what has been described so far, some applications to Bessel functions are discussed in this section as an extension of their brief introduction in Sect. 2.1. The Bessel[5] function appears as the solution of Bessel's equation after a cylindrical polar coordinate decomposition of the Laplace equation (see Chap. 1 of [31]). These functions are typical of the type encountered in applied sciences and are realistic examples in that they combine both exponential and oscillatory behavior on the interval of definition, $r \in [0, \infty]$. Therefore, they are suitable for application of numerical methods developed in this book.

The Bessel function $J_\mu(r)$, for non-integer values of μ is defined by the infinite power series

$$J_\mu(r) = \left(\frac{r}{2}\right)^\mu \sum_{k=0}^{\infty} c(\mu, k) r^{2k}, \qquad (2.88)$$

where the coefficient $c(\mu, k)$ is defined as

$$c(\mu, k) = \frac{(-1)^k}{2^{2k} k! \Gamma(\mu + k + 1)}, \qquad (2.89)$$

and the Gamma function in (2.89) is defined by the integral

$$\Gamma(z) = \int_0^\infty e^{-x} x^{z-1} dx, \qquad (2.90)$$

that is convergent for $z > 0$ and has the recurrence property

$$\Gamma(z+1) = z \Gamma(z). \qquad (2.91)$$

For the case that z is integer

$$\Gamma(z+1) = z!, \qquad (2.92)$$

while for the half-integer case

$$\Gamma\left(\frac{1}{2}\right) = \sqrt{\pi}, \qquad (2.93)$$

and (2.91) applies, so that $\Gamma(\frac{3}{2}) = \frac{1}{2}\Gamma(\frac{1}{2})$ and so on.

---

[4] Baron (Jean Baptiste) Joseph Fourier, French mathematician 1768–1830.
[5] Friedrich Wilhelm Bessel, German astronomer 1784–1846.

2.6 Application to Bessel Functions

The Bessel function $J_\mu(r)$ satisfies the three-term recurrence

$$J_{\mu-1}(r) = \frac{2\mu}{r} J_\mu(r) - J_{\mu+1}(r), \tag{2.94}$$

and similar results for the derivative with respect to argument

$$2\frac{dJ_\mu(r)}{dr} = J_{\mu-1}(r) - J_{\mu+1}(r), \tag{2.95}$$

$$\frac{dJ_\mu(r)}{dr} = \frac{\mu}{r} J_\mu(r) - J_{\mu+1}(r). \tag{2.96}$$

These recurrences are stable in numerical applications when applied in decreasing order for a fixed argument r. Any two values may be used to start the downward recurrence in (2.94), but it is appropriate to commence with two starting values obtained from the asymptotic expression for large orders

$$J_\mu(r) \sim \frac{1}{\sqrt{2\pi\mu}} \left(\frac{er}{2\mu}\right)^\mu. \tag{2.97}$$

Of special interest are the cases when μ is either an integer, $\mu = \ell$, or half-integer $\mu = \ell + \frac{1}{2}$. In the former case $J_\ell(r)$ is the Bessel function of integer order and in the latter case, the spherical Bessel function $j_\ell(r)$ is related to the half-integer order Bessel function by

$$j_\ell(r) = \sqrt{\frac{\pi}{2r}} J_{\ell+\frac{1}{2}}(r), \tag{2.98}$$

with the asymptotic expression for large order

$$j_\ell(r) \sim \sqrt{\frac{e}{2}} \frac{(er)^\ell}{(2\ell+1)^{\ell+1}}. \tag{2.99}$$

Recurrence relations for $j_\ell(r)$ follow from substitution of $\mu = \ell + \frac{1}{2}$ into any of (2.94)–(2.96) to give the following result

$$j_{\ell-1}(r) = \frac{2\ell+1}{r} j_\ell(r) - j_{\ell+1}(r), \tag{2.100}$$

and a similar result for the derivative with respect to argument

$$(2\ell+1)\frac{dj_\ell(r)}{dr} = \ell j_{\ell-1}(r) - (\ell+1) j_{\ell+1}(r). \tag{2.101}$$

The sequence of values generated from any of these recurrences, down to the lowest order, differs from the exact value by an undetermined constant. The exact values must be obtained by normalizing the sequence and this is accomplished by using a further property of these functions. The Bessel functions of integer order $J_\ell(r)$

satisfy the summation formula

$$1 = J_0^2(r) + 2\sum_{\ell=1}^{\infty} J_\ell^2(r), \qquad (2.102)$$

while those for half-integer order have the summation expression

$$1 = \sum_{\ell=0}^{\infty} (2\ell+1) j_\ell^2(r). \qquad (2.103)$$

When the sequence obtained from recurrence is substituted into either of (2.102) or (2.103), then the result on the left-hand side is not unity, but some other number. The square root of this number, when divided into the sequence obtained from recurrence, produces the normalized values of the respective functions. After normalization it is important to compare the accuracy of the recurrence values against those of the exact expressions for the lowest orders. As an example, the first few spherical Bessel functions have the explicit expressions

$$j_0(r) = \frac{sin(r)}{r}, \qquad (2.104)$$

$$j_1(r) = \frac{1}{r}\left[\frac{sin(r)}{r} - cos(r)\right], \qquad (2.105)$$

$$j_2(r) = \frac{1}{r}\left[\left(\frac{3}{r^2} - 1\right) sin(r) - \frac{3}{r} cos(r)\right], \qquad (2.106)$$

and they satisfy the limit in argument $r \to 0$

$$\frac{j_\ell(r)}{r^\ell} \to \frac{1}{1 \cdot 3 \cdot 5 \cdots (2\ell+1)}. \qquad (2.107)$$

The spherical Bessel function satisfies the second-order differential equation

$$r^2 \frac{d^2 j_\ell(r)}{dr^2} + 2r \frac{d j_\ell(r)}{dr} + [r^2 - \ell(\ell+1)] j_\ell(r) = 0. \qquad (2.108)$$

If $r \in [0, a]$, on application of the mapping

$$r = ax, \qquad (2.109)$$

with $x \in [0, +1]$, an expansion in a series of the even Chebyshev polynomials is defined by

$$j_\ell(ax) = \frac{\sqrt{\pi}}{2}\left(\frac{ax}{2}\right)^\ell \sum_{n=0}^{N\prime} B_n^\ell(a) T_{2n}(x), \qquad (2.110)$$

and N denotes the value $n = N$, at which the series is truncated for the approximation to the left-hand side. Note that the coefficients of the expansion in (2.110) depend

2.6 Application to Bessel Functions

on the value of a, if the range of r changes, then so do the coefficients. On noting the property

$$x^\ell T_{2n}(x) = \frac{1}{2^\ell} \sum_{k=0}^{\ell} \binom{\ell}{k} T_{|2n+\ell-2k|}(x), \qquad (2.111)$$

then (2.110) may be written

$$j_\ell(ax) = \sum_{i=0}^{N+m} {}' C_i^\ell(a) T_{2i+p}(x), \qquad (2.112)$$

with

$$m = \frac{\ell - p}{2}, \qquad (2.113)$$

$$p = \mathrm{mod}(\ell, 2), \qquad (2.114)$$

and, after a little algebra, it may be shown that

$$C_i^\ell(a) = \frac{\sqrt{\pi}}{2} \frac{a^\ell}{2^{2\ell}} \left[\sum_{n=n_1}^{n_2} \binom{\ell}{n-i+m} B_n^\ell(a) + \gamma_i \sum_{n=0}^{n_3} \binom{\ell}{n+i+m+p} B_n^\ell(a) \right], \qquad (2.115)$$

where

$$\gamma_i = 1, \ i \le m,$$
$$= 0, \ i > m,$$
$$n_1 = \max(0, i - m),$$
$$n_2 = \min(N, i + m + p), \qquad (2.116)$$
$$n_3 = \min(N, m - i).$$

Explicit expressions, as well as a recurrence relation for the coefficients $B_n^\ell(a)$ of (2.110), have been given by Luke [14] in Sect. 9.3.6. The following is a brief description of the author's publication of a Fortran code to compute the Chebyshev coefficients [32] where more details of this code may be found.

Introducing the notation

$$\epsilon_n = 1, \ n = 0,$$
$$= 2, \ n > 0. \qquad (2.117)$$

The recurrence relation for $B_n^\ell(a)$ is

$$\frac{2B_n^\ell(a)}{\epsilon_n} = -\left[\frac{16(n+1)(n+\ell+\frac{3}{2})}{a^2} - \frac{(n+1)}{(n+2)}\right] B_{n+1}^\ell(a)$$
$$- \left[\frac{16(n+1)(n-\ell+\frac{3}{2})}{a^2} - 1\right] B_{n+2}^\ell(a)$$
$$- \frac{(n+1)}{(n+2)} B_{n+3}^\ell(a). \tag{2.118}$$

This downward recurrence is convergent (see Sect. 16.2 of [14]) and because, $T_{2n}(0) = (-1)^n$, from (2.107) the normalization condition is

$$\sum_{n=0}^{N'}(-1)^n B_n^\ell(a) = \frac{1}{\sqrt{\pi}} \frac{2^{\ell+1}}{1 \cdot 3 \cdot 5 \cdots (2\ell+1)}. \tag{2.119}$$

From the recurrence relations for $T_n(x)$ and $j_\ell(r)$, those for the coefficients $B_n^\ell(a)$ and $C_i^\ell(a)$ may be shown to be

$$B_n^{\ell-1}(a) = \left(\frac{2\ell+1}{2}\right) B_n^\ell(a) - \left(\frac{a}{4}\right)^2 \left[B_{n+1}^{\ell+1}(a) + 2B_n^{\ell+1}(a) + B_{|n-1|}^{\ell+1}(a)\right], \tag{2.120}$$

and

$$C_{|i-1+p|}^{\ell-1}(a) = \frac{2(2\ell+1)}{a} C_i^\ell(a) - C_{i+p}^{\ell+1}(a) - C_{|i-1+p|}^{\ell+1}(a) - C_{i+p}^{\ell-1}(a). \tag{2.121}$$

The Chebyshev series of (2.110) and (2.112) may then be evaluated by the recurrence relations for even and odd series of Chebyshev polynomials, as discussed for (2.66) to (2.70). The accompanying Fortran code, JLRCHB, computes the coefficients in the last two equations, values from the series expansions in Chebyshev polynomials and compares results with those obtained from downward recurrence in (2.100). Details may be found in [32] and commentary in the code itself.

Several of these expressions given in this section have been applied in examples, where some graphs of these functions are shown. The monomial series sum of (2.88) is computed for the spherical Bessel function case. However, monomial series summation is not the most efficient method of approximating these functions as convergence may be slow. Therefore, improved techniques used in the Fortran code are preferred.

EXAMPLES 2.19–2.23
FORTRAN CODE JLRCHB

2.7 Application to the Associated Legendre Polynomial

The Associated Legendre polynomial arises in classical potential theory and appears in the spherical polar coordinate decomposition of Laplace's equation (see Chap. 3 of [31]) as the solution of the Legendre equation (see Chap. 24 of [8] where it is called the Associated Legendre function). It also appears in nuclear scattering computations where methods for evaluation to high order and precision were required in Fortran code developed by the author [33–36]. This section briefly describes the Chebyshev expansion of the Associated Legendre polynomial with the author's Fortran code published in [37]. The method used is due to Clenshaw [18] and the coefficients of the expansion need only be computed once, then stored in tabular form to enable the most efficient performance in applications.

The Associated Legendre polynomial, $P_\ell^m(x)$, with m, ℓ integer, satisfies the second-order differential equation

$$(x^2 - 1)\frac{d^2 y}{dx^2} + 2x\frac{dy}{dx} - \left[\ell(\ell+1) - \frac{m^2}{(1-x^2)}\right]y = 0, \quad -1 \leq x \leq +1. \quad (2.122)$$

These have the property

$$P_\ell^m(-x) = (-1)^{\ell-m} P_\ell^m(x), \quad (2.123)$$

and therefore only values of x in the positive interval need to be considered. The boundary conditions are (page 145, of [38])

$$P_\ell^m(1) = 1, \ m = 0,$$
$$= 0, \ m > 0, \quad (2.124)$$

$$P_\ell^m(0) = (-1)^m \frac{2^m}{\sqrt{\pi}} \cos\left[\frac{\pi}{2}(m+\ell)\right] \frac{\Gamma\left(\frac{\ell+m+1}{2}\right)}{\Gamma\left(\frac{\ell-m+2}{2}\right)}, \ m+\ell \text{ even},$$
$$= 0, \ m+\ell \text{ odd}. \quad (2.125)$$

For the special case of $m = \ell$

$$P_\ell^\ell(x) = \frac{(2\ell)!}{2^\ell \ell!}(1-x^2)^{\frac{\ell}{2}}, \quad (2.126)$$

while, for $m = 0$, the usual Legendre polynomial, $P_\ell(x)$ (defined in Sect. 2.4) is obtained. From the definition

$$P_\ell^m(x) = (1-x^2)^{\frac{m}{2}} \frac{d^m P_\ell(x)}{dx^m}, \quad (2.127)$$

it is seen that, excepting the case when m is odd, $P_\ell^m(x)$ is a polynomial in odd or even powers of x as ℓ is odd or even. Therefore, the following Chebyshev expansions are introduced (for m even only)

$$P_\ell^m(x) = \sum_{j=0}^{\frac{\ell}{2}{'}} a_j^{lm} T_{2j}(x), \quad \ell \text{ even}, \tag{2.128}$$

$$P_\ell^m(x) = \sum_{j=0}^{\frac{(\ell-1)}{2}} a_j^{lm} T_{2j+1}(x), \quad \ell \text{ odd}, \tag{2.129}$$

and both expansions are finite sums (note the prime on the first of these). For the case, $m = 0$, i.e., the Legendre polynomial, the expansion coefficients in both expansions have simple analytical expressions [39]. In the case that m is odd, the factor $(1-x^2)^{\frac{1}{2}}$ has the Chebyshev series ([5], p. 135, exercise 3.4.1(d))

$$(1 - x^2)^{\frac{1}{2}} = \frac{2}{\pi} - \frac{4}{\pi} \sum_{j=1}^{\infty} \frac{1}{4j^2 - 1} T_{2j}(x), \tag{2.130}$$

which is infinite and poorly convergent, so recurrence relations are a suitable alternative.

One recurrence scheme for generating $P_\ell^m(x)$ would be to start with a recurrence in ℓ, for $m = 0$, using as initial values, $P_0(x) = 1$, $P_1(x) = x$, and on rearranging (2.43), to obtain

$$P_\ell^0(x) = \frac{1}{\ell} \left[(2\ell - 1) x P_{\ell-1}^0(x) - (\ell - 1) P_{\ell-2}^0(x) \right], \tag{2.131}$$

next apply recurrence in ℓ and m,

$$P_\ell^{m+1}(x) = (1 - x^2)^{-\frac{1}{2}} \left[(\ell + m) P_{\ell-1}^m(x) - (\ell - m) x P_\ell^m(x) \right], \tag{2.132}$$

and then a recurrence in m,

$$P_\ell^{m+2}(x) = 2(m+1) x (1 - x^2)^{-\frac{1}{2}} P_\ell^{m+1}(x) - (\ell + m + 1)(\ell - m) P_\ell^m(x). \tag{2.133}$$

Alternatively, following the method of Clenshaw [18] and substituting (2.128) or (2.129) into the differential equation (2.122) satisfied by the Associated Legendre polynomial, the resulting three-term recurrence scheme for the coefficients a_j^{lm}, follows after a little algebra.

For ℓ even: $a_{\ell/2+1}^{lm} = 0$, $a_{\ell/2}^{lm} = 1$, the recurrence is

$$[2j(2j+1) - \ell(\ell+1)] a_j^{lm} = 2 \left[(2j+2)^2 - \ell(\ell+1) + 2m^2 \right] a_{j+1}^{lm}$$
$$+ [\ell(\ell+1) - (2j+4)(2j+3)] a_{j+2}^{lm}, \tag{2.134}$$

2.7 Application to the Associated Legendre Polynomial

Table 2.5 Results for P_ℓ^m

Absolute error	Maximum ℓ	Maximum m
10^{-10}	> 100	> 10
10^{-15}	26	8
10^{-20}	10	6
10^{-25}	6	6

while for ℓ odd: $a^{\ell m}_{(\ell+1)/2} = 0$, $a^{\ell m}_{(\ell-1)/2} = 1$, it is

$$[(2j+1)(2j+2) - \ell(\ell+1)]a^{\ell m}_j = 2\left[(2j+3)^2 - \ell(\ell+1) + 2m^2\right]a^{\ell m}_{j+1}$$
$$+ [\ell(\ell+1) - (2j+5)(2j+4)]a^{\ell m}_{j+2}. \quad (2.135)$$

The coefficients are normalized using the boundary condition at $x = 0$, when $m + \ell$ is even. For m even and ℓ odd, the general form of the recurrence (2.131) may be used

$$xP_\ell^m(x) = \left(\frac{\ell-m+1}{2\ell+1}\right)P_{\ell+1}^m(x) + \left(\frac{\ell+m}{2\ell+1}\right)P_{\ell-1}^m(x), \quad (2.136)$$

where, upon substituting for $P_{\ell+1}^m(x)$ and $P_{\ell-1}^m(x)$, in the appropriate expansion (2.128), (2.129), the following relation for coefficients is obtained

$$a^{\ell m}_j = \left[\frac{2(\ell-m+1)}{2\ell+1}\right]a^{\ell+1m}_j + \left[\frac{2(\ell+m)}{2\ell+1}\right]a^{\ell-1m}_j - a^{\ell m}_{j-1} \quad (2.137)$$

with j in the range 0 to $(\ell-1)/2$ and $a^{\ell m}_{-1} = 0$. The Chebyshev expansions of (2.128), (2.129) may then be evaluated by recurrence, as in (2.66) to (2.70). For the case that m is odd, this method requires the prior calculation of $P_\ell^{m\pm 1}$, and application of (2.133).

The accompanying Fortran package, PLMCHB, computes the coefficients in the last equation, the values from the series expansions in Chebyshev polynomials and compares the result to values obtained from recurrence relations for P_l^m. Table 2.5 summarizes results for the maximum absolute error allowed in comparing the two methods for a range in argument. Details on precision, accuracy and error may be found in [37].

EXAMPLE 2.24
FORTRAN CODE PLMCHB

2.8 Application to the Parabolic Cylinder Function

A theoretical nuclear physics application [40] called for the computation of the parabolic cylinder function and integrals of products of such functions (see Chap. 46 of [11] and Chap. 19 of [6]). For this purpose, a Fortran code, DLRCHB, was developed by the author using Chebyshev series expansions of the function and the first-order derivative in argument and order. The method has limitations, largely because of the asymptotic behavior of the parabolic cylinder function, but has important computational advantages in applications demonstrated in two examples from [40].

The parabolic cylinder function $D_\lambda(z)$ is the solution of the differential equation (Weber's equation)

$$\frac{d^2 D_\lambda(z)}{dz^2} + \left(\lambda + \frac{1}{2} - \frac{z^2}{4}\right) D_\lambda(z) = 0, \tag{2.138}$$

with λ, z continuous variables that may be complex. Expressed in terms of confluent hypergeometric functions, $D_\lambda(z)$, is defined as

$$D_\lambda(z) = 2^{\frac{\lambda}{2}} e^{\frac{-z^2}{4}} \left[\frac{\Gamma(\frac{1}{2})}{\Gamma(\frac{1-\lambda}{2})} \,_1F_1\left(-\frac{\lambda}{2}; \frac{1}{2}; \frac{z^2}{2}\right) + \frac{z}{\sqrt{2}} \frac{\Gamma(-\frac{1}{2})}{\Gamma(-\frac{\lambda}{2})} \,_1F_1\left(\frac{1-\lambda}{2}; \frac{3}{2}; \frac{z^2}{2}\right) \right]. \tag{2.139}$$

where Γ is the usual gamma function (see Sect. 2.1). The expressions of (2.138), (2.139) may be found in Sect. 8.2 of [41] together with other properties. Luke, in Chap. V of [21], presents expansions in shifted Chebyshev polynomials for the confluent hypergeometric function [38,42] $_1F_1(a; c; z)$ where the argument z differs from that in (2.139). At the origin, $z = 0$, the values of $D_\lambda(z)$ and its first derivative, with respect to argument for λ integer, are

$$D_\lambda(0) = \frac{\Gamma(\frac{1}{2}) 2^{\frac{\lambda}{2}}}{\Gamma(\frac{1-\lambda}{2})}, \tag{2.140}$$

$$D'_\lambda(0) = \frac{\Gamma(-\frac{1}{2}) 2^{\frac{\lambda-1}{2}}}{\Gamma(-\frac{\lambda}{2})}. \tag{2.141}$$

For large z asymptotic expansions for $D_\lambda(z)$ in various regions of the complex plane are given in [43], where it is seen that for z real and positive

$$e^{\frac{z^2}{4}} D_\lambda(z) \to z^\lambda \times \left[\text{an asymptotic series in } z^{-2k} \right]. \tag{2.142}$$

It is the possibility of fractional powers of λ that limits the method described here when z is sufficiently large. If λ is non-integral then, $D_\lambda(z)$ and $D_\lambda(-z)$ are linearly independent solutions of (2.138) and satisfy the Wronskian relation

$$D_\lambda(z) D'_\lambda(-z) - D_\lambda(-z) D'_\lambda(z) = \frac{\sqrt{2\pi}}{\Gamma(-\lambda)}, \tag{2.143}$$

2.8 Application to the Parabolic Cylinder Function

where the prime denotes differentiation with respect to the argument z. If $\lambda = k$, an integer, then the parabolic cylinder function is related to the Hermite polynomial (see Chap. 22 of [6])

$$D_k(z) = 2^{-\frac{k}{2}} e^{\frac{-z^2}{4}} H_k\left(\frac{z}{\sqrt{2}}\right), \tag{2.144}$$

or, with the change of variable $z = \sqrt{2}\,r$,

$$D_k(\sqrt{2}\,r) = 2^{-\frac{k}{2}} e^{\frac{-r^2}{2}} H_k(r). \tag{2.145}$$

On introducing the variable $x \in [-1, +1]$, such that $z = \sqrt{2}ax$, $z \in [-\sqrt{2}a, +\sqrt{2}a]$, then a Chebyshev series may be defined on this interval as

$$D_\lambda(\sqrt{2}\,r) = 2^{-\frac{\lambda}{2}} e^{\frac{-a^2 x^2}{2}} \sum_{n=0}^{\infty'} B_n^\lambda(a) T_n(x), \tag{2.146}$$

where the Chebyshev polynomials are orthogonal on the interval $x = z/\sqrt{2}a \in [-1, +1]$. The prime on the series indicates that the first term contains a factor of $\frac{1}{2}$. In the case that $\lambda = k$, as in (2.145), the Chebyshev series terminates at $n = k$, because it is the expansion of a polynomial in x.

The differential equation (2.138) has coefficients that are simple polynomials in z and therefore, following the method of Clenshaw [18], a simple recurrence in n for the $B_n^\lambda(a)$ coefficients follows if (2.146) is substituted into (2.138). Before proceeding with this, some additional definitions are introduced.

For the first derivative of $D_\lambda(\sqrt{2}\,r)$ with respect to argument z, introduce the series

$$D'_\lambda(\sqrt{2}\,r) = 2^{-\frac{\lambda}{2}} e^{\frac{-a^2 x^2}{2}} \sum_{n=0}^{\infty'} d_n^\lambda(a) T_n(x), \tag{2.147}$$

while for the first derivative with respect to order of $D_\lambda(\sqrt{2}\,r)$

$$\frac{\partial D_\lambda(\sqrt{2}\,r)}{\partial \lambda} = -\frac{\ln(2)}{2} D_\lambda(\sqrt{2}\,r) + 2^{-\frac{\lambda}{2}} e^{\frac{-a^2 x^2}{2}} \sum_{n=0}^{\infty'} \frac{\partial B_n^\lambda(a)}{\partial \lambda} T_n(x), \tag{2.148}$$

and for the derivative with respect to argument and order

$$\frac{\partial D'_\lambda(\sqrt{2}\,r)}{\partial \lambda} = -\frac{\ln(2)}{2} D'_\lambda(\sqrt{2}\,r) + 2^{-\frac{\lambda}{2}} e^{\frac{-a^2 x^2}{2}} \sum_{n=0}^{\infty'} \frac{\partial d_n^\lambda(a)}{\partial \lambda} T_n(x). \tag{2.149}$$

Note the order of differentiation in (2.149); reversing the order of differentiation corresponds to differentiating (2.148) with respect to argument z and this leads to a different second term in (2.149).

A recurrence relation for the coefficients $B_n^\lambda(a)$ follows on substitution of the series (2.146) into (2.138) and applying the orthogonality relation (2.48) after multiplication by $T_n(x)$.

$$(n+3)(\lambda-n)B_n^\lambda(a) = 2(n+2)\left[\lambda+1-\frac{(n+1)(n+3)}{a^2}\right]B_{n+2}^\lambda(a)$$
$$-(n+1)(\lambda+n+4)B_{n+4}^\lambda(a). \quad (2.150)$$

For integer $\lambda = k$, the series in (2.146) terminates at $n = k$, while for non-integer values of λ, the function value is approximated by truncating the series at some value of $n = N + 2$. For both odd and even values of n in (2.150) the recurrence is commenced with $B_{N+4}^\lambda(a) = 0$, $B_{N+2}^\lambda(a) = 1$, and the recurrence is specific for each λ and a.

The series in (2.146) has both odd and even terms and from values of $T_k(0)$ in Table 2.3, it is seen that the boundary condition (2.140) normalizes only the even coefficients. The normalization condition for the odd coefficients follows on differentiating (2.146) with respect to argument. On noting (2.46), (2.48), applying the second boundary condition (2.141) and using properties of the gamma function, the following two normalization conditions are obtained

$$\sum_{m=0}^{\frac{N}{2}\prime}(-1)^m B_{2m}^\lambda(a) = \left(\frac{2^\lambda}{\sqrt{\pi}}\right)\Gamma\left(\frac{1+\lambda}{2}\right)\cos\left(\frac{\pi\lambda}{2}\right), \quad (2.151)$$

$$\sum_{m=0}^{\frac{N}{2}-1}(-1)^m(2m+1)B_{2m+1}^\lambda(a) = \left(\frac{a2^{\lambda+1}}{\sqrt{\pi}}\right)\Gamma\left(1+\frac{\lambda}{2}\right)\sin\left(\frac{\pi\lambda}{2}\right). \quad (2.152)$$

An expression for the coefficients d_n^λ in (2.147) follows on differentiating (2.146) with respect to argument z, and equating coefficients of $T_n(x)$

$$d_n^\lambda(a) = d_{n+2}^\lambda(a)$$
$$-\frac{a}{2\sqrt{2}}\left[B_{|n-1|}^\lambda(a) - \frac{4(n+1)}{a^2}B_{n+1}^\lambda(a) - B_{n+3}^\lambda(a)\right]. \quad (2.153)$$

A recurrence for the coefficients $B_n^{\lambda\prime}(a) = \partial B_n^\lambda(a)/\partial\lambda$ in (2.148) follows on differentiating the recurrence (2.150) with respect to λ

$$(n+3)\left[(\lambda-n)B_n^{\lambda\prime}(a) + B_n^\lambda(a)\right] = 2(n+2) \times$$
$$\left[\left\{\lambda+1-\frac{(n+1)(n+3)}{a^2}\right\}B_{n+2}^{\lambda\prime}(a) + B_{n+2}^\lambda(a)\right]$$
$$-(n+1)\left[(\lambda+n+4)B_{n+4}^{\lambda\prime}(a) + B_{n+4}^\lambda(a)\right]. \quad (2.154)$$

The last two recurrence relations use coefficients $B_n^\lambda(a)$ that are normalized as in (2.151), (2.152) and therefore the normalized values of $d_n^\lambda(a)$ and $B_n^{\lambda\prime}(a)$ are obtained.

A recurrence for the coefficients $d_n^{\lambda\prime}(a) = \partial d_n^\lambda(a)/\partial\lambda$ in (2.149) follows on differentiating (2.153) with respect to λ, i.e., replacing $B_n^\lambda(a)$ by $B_n^{\lambda\prime}(a)$ and $d_n^\lambda(a)$ by $d_n^{\lambda\prime}(a)$.

To evaluate Chebyshev series values in (2.146) and (2.147), coefficients are generated by the FORTRAN CODE DLRCHB and then used as input for the examples. The download of this Fortran code includes a copy of an unpublished report on which the above exposition is based [44]. This report provides a description of the included subroutines, examples of applications and accuracy issues.

Comparisons of the numerical results were made against tabulated values in Chap. 19 of Abramowitz and Stegun [6]. However, because these are limited to 5 significant figures, a more accurate comparison was made against values generated from the Fortran code listed by Luke (Chap. V in [21]). This is included as FORTRAN CODE DLRLUK and is applied twice to generate the Chebyshev coefficients for the two terms in (2.139). However, note needs to be made of Luke's use of a scaling factor $z = wx$, where z is the third argument in the confluent hypergeometric function $_1F_1(a; c; z)$ that also differs from the usage in (2.139). Thus a change in variable is required if the definition of the third argument in (2.139) is applied where the scaling factor chosen here is $w = 9 = a^2$. This last comparison shows outstanding agreement in precision. For differences in the recurrence relations for the different Chebyshev polynomials (shifted versus non-shifted, etc), see the summary table at the end of Chap. VIII in Volume I of [14].

EXAMPLES 2.25–2.28
FORTRAN CODES DLRCHB, DLRLUK

2.9 Convolution of Series Products

In applications it is often necessary to compute the product of two (or more) polynomial series. When series are either of the Chebyshev, or trigonometric type, special properties reduce the product to a single series. Thus, two Chebyshev series, each consisting of $N + 1$ terms, when multiplied together, can be converted to a single Chebyshev series of $2N+1$ terms. For functions $A(t)$, $B(t)$, defined on $t \in [-1, +1]$, consider the convolution of two Chebyshev series

$$A(t) = \sum_{k=0}^{\infty} \frac{1}{2} \epsilon_k a_k T_k(t), \qquad (2.155)$$

$$B(t) = \sum_{k=0}^{\infty} \frac{1}{2} \epsilon_k b_k T_k(t), \qquad (2.156)$$

the product is

$$C(t) = A(t)B(t),$$
$$= \sum_{k=0}^{\infty} \frac{1}{2} \epsilon_k c_k T_k(t), \qquad (2.157)$$

where $\epsilon_0 = 1$, $\epsilon_k = 2$, $k \geq 1$, is used in place of the prime on the summation. Expressions for the coefficients c_k follow from the Chebyshev polynomial property (p. 5 of [5])

$$T_m(t)T_n(t) = \frac{1}{2}T_{m+n}(t) + \frac{1}{2}T_{|m-n|}(t), \qquad (2.158)$$

that gives for $k = 0$

$$c_0 = \frac{1}{2}a_0 b_0 + \sum_{j=1}^{\infty} a_j b_j, \qquad (2.159)$$

and for $k > 0$

$$c_k = \frac{1}{2}a_0 b_k + \frac{1}{2}\sum_{j=1}^{\infty} a_j \left[b_{|k-j|} + b_{k+j} \right], \qquad (2.160)$$

$$= \frac{1}{2}b_0 a_k + \frac{1}{2}\sum_{j=1}^{\infty} b_j \left[a_{|k-j|} + a_{k+j} \right]. \qquad (2.161)$$

These expressions differ from the ones given by Luke (see Sect. 8.6.1 of [14]) because there they depart from the convention that defines the first Chebyshev coefficient written as $\frac{1}{2}a_0$. The convolution results above are unchanged in the case of the shifted Chebyshev polynomials (see Sect. 3.8 of [4]).

When applied to series (2.155), (2.156) that terminate at $j = N$, the summations in (2.159)–(2.161) are also truncated at $j = N$. But serious errors are introduced if the summations of (2.160) or (2.161) are truncated at $\frac{1}{2}N$. To avoid this problem, a simple scheme is to set to zero input coefficients a_j, b_j for $j = N+1, \ldots, 2N$. Then the product series in (2.157) is usually truncated at the same length as the series in (2.155), (2.156), namely, $N+1$ terms. However, this often introduces a truncation error that leads to errors that are larger in the product series for $C(t)$, than those for the separate series of $A(t)$, or $B(t)$. Therefore, caution is appropriate when the convolution product is applied repeatedly.

Similarly to the Chebyshev case, two Fourier series, each consisting of $N+1$ terms, when multiplied together, can be converted to a single Fourier series of $2N+1$ terms in a convolution product. Consider two Fourier series

$$A(t) = \sum_{k=0}^{\infty} \frac{1}{2}\epsilon_k c'_k \cos(k\pi t) + \sum_{k=1}^{\infty} s'_k \sin(k\pi t), \qquad (2.162)$$

$$B(t) = \sum_{k=0}^{\infty} \frac{1}{2}\epsilon_k c''_k \cos(k\pi t) + \sum_{k=1}^{\infty} s''_k \sin(k\pi t), \qquad (2.163)$$

2.9 Convolution of Series Products

that have the product

$$C(t) = A(t)B(t) \tag{2.164}$$

$$= \sum_{k=0}^{\infty} \frac{1}{2}\epsilon_k c_k \cos(k\pi t) + \sum_{k=1}^{\infty} s_k \sin(k\pi t), \tag{2.165}$$

where again $\epsilon_0 = 1$, $\epsilon_k = 2$, $k \geq 1$, is used in place of a prime on the summation. Expressions for the coefficients c_k, s_k, follow from application of the orthogonality properties in (2.75)–(2.77) and the standard relations for the trigonometric functions

$$\cos(m\theta)\cos(n\theta) = \frac{1}{2}\cos\left[(m+n)\theta\right] + \frac{1}{2}\cos\left[(m-n)\theta\right], \tag{2.166}$$

$$\sin(m\theta)\sin(n\theta) = \frac{1}{2}\cos\left[(m-n)\theta\right] - \frac{1}{2}\cos\left[(m+n)\theta\right], \tag{2.167}$$

$$\cos(m\theta)\sin(n\theta) = \frac{1}{2}\sin\left[(m+n)\theta\right] - \frac{1}{2}\cos\left[(m-n)\theta\right]. \tag{2.168}$$

This gives the result for $k = 0$

$$c_0 = \frac{1}{2}c_0' c_0'' + \sum_{j=1}^{\infty} \left(c_j' c_j'' + s_j' s_j''\right), \tag{2.169}$$

and for $k > 0$ the result is

$$\begin{aligned}
c_k =& \frac{1}{2}c_0' c_k'' + \frac{1}{2}c_k' c_0'' \\
&+ \frac{1}{2}\sum_{j=1}^{\infty} \left(c_j' c_{j+k}'' + s_j' s_{j+k}''\right) \\
&+ \frac{1}{2}\sum_{j=1}^{k-1} \left(c_j' c_{k-j}'' - s_j' s_{k-j}''\right) \\
&+ \frac{1}{2}\sum_{j=k+1}^{\infty} \left(c_j' c_{j-k}'' + s_j' s_{j-k}''\right),
\end{aligned} \tag{2.170}$$

$$\begin{aligned}
s_k =& \frac{1}{2}c_0' s_k'' + \frac{1}{2}s_k' c_0'' \\
&+ \frac{1}{2}\sum_{j=1}^{\infty} \left(c_j' s_{j+k}'' - s_j' c_{j+k}''\right) \\
&+ \frac{1}{2}\sum_{j=1}^{k-1} \left(c_j' s_{k-j}'' + s_j' c_{k-j}''\right)
\end{aligned}$$

$$-\frac{1}{2}\sum_{j=k+1}^{\infty}\left(c'_j s''_{j-k} - s'_j c''_{j-k}\right), \qquad (2.171)$$

with the summations $j = 1$ to $k - 1$ being absent if $k = 1$, and $s'_0 = s''_0 = 0$.

Fortran code for calculation of the coefficients for convolution of two Chebyshev or Fourier series has been published by the author [45] and some applications are the subject of Sect. 4.8.

EXAMPLE 2.29
FORTRAN CODE CCFS

2.10 Types of Convergence

This chapter is concluded with a brief discussion of convergence characteristics for series of the form in (2.13). Much more is said on this topic in Chap. 4, once some analytical background material has been studied, but some simple results are introduced here. Obviously, the series representation in (2.13), of an arbitrary function, can only be used in truncated form as an approximation, if the subsequent terms in the series decrease in magnitude. However, not only should the magnitude decrease, but it should do so at a rate that ensures a minimal number of terms in the approximation. The actual number of terms required for a specific accuracy will depend on factors such as the smoothness behavior of the function being approximated, the size of the interval of approximation, $[a, b]$, and the choice of basis set $\{\psi_K\}_1^\infty$. Each of these issues is investigated separately in subsequent chapters. In this section, some simple methods of estimating the effects of truncating the series in (2.13) are described.

Given that a series representation for the function $f(x)$ exists on $x \in [-1, +1]$ (for simplicity), with a basis, $p_{K-1}(x) = \psi_K(x)$, consider the truncation of the series in (2.13) that is defined by

$$f_N(x) = \sum_{k=0}^{N} a_k p_k(x). \qquad (2.172)$$

To determine estimates for the error of the truncation in (2.172), consider the error curve of (2.15), defined here as

$$E_N(x) = f(x) - f_N(x). \qquad (2.173)$$

Two related aspects of the approximation in (2.172) to $f(x)$ need to be considered. First of these is an estimate for the bound on $E_N(x)$ in the interval $x \in [-1, +1]$, and second is the rate of decrease of the coefficients, a_k, in the limit $k \to \infty$. The discussion is simplified if the choice of basis set $\{p_k\}_0^\infty$ is restricted to the Legendre and Chebyshev polynomials, or the Fourier basis, because a bound on these basis elements satisfies $|p_k(x)| \le 1$. To estimate the bound on $E_N(x)$, consider again (2.173)

2.10 Types of Convergence

$$|E_N(x)| = |f(x) - f_N(x)|,$$

$$= \left|\sum_{k=N+1}^{\infty} a_k p_k(x)\right|,$$

$$\leq \sum_{k=N+1}^{\infty} |a_k| |p_k(x)|,$$

$$\leq \sum_{k=N+1}^{\infty} |a_k|. \tag{2.174}$$

Therefore, if the series is convergent, the bound in (2.174) should establish if the sequence of numbers $|E_0(x)|, |E_1(x)|, |E_2(x)|, \ldots$, is convergent such that

$$\lim_{N \to \infty} |E_N(x)| \to 0. \tag{2.175}$$

Therefore, the rate of decrease of this sequence is related to the second aspect of the approximation (2.172), namely, the rate of decrease in the magnitude of the coefficient a_k. To investigate this behavior consider the example of the Fourier series for the exponential function, $f(x) = e^x$, on the interval $x \in [-1, +1]$, for which application of (2.73), (2.74) and integration by parts gives the analytical result for the cosine and sine series coefficients

$$c_k = \int_{-1}^{+1} e^x \cos(k\pi x) dx, \quad k = 0, 1, 2, 3, \ldots, \tag{2.176}$$

$$s_k = \int_{-1}^{+1} e^x \sin(k\pi x) dx, \quad k = 1, 2, 3, \ldots, \tag{2.177}$$

and integration by parts (p. 451 of [46]), gives

$$c_k = (-1)^k \left[\frac{e - e^{-1}}{1 + (k\pi)^2}\right], \quad k = 0, 1, 2, 3, \ldots, \tag{2.178}$$

$$s_k = -k\pi c_k, \quad k = 1, 2, 3, \ldots. \tag{2.179}$$

Thus, the rate of decrease of the two sets of coefficients in the limit $k \to \infty$, is

$$c_k \sim \mathcal{O}(k^{-2}), \tag{2.180}$$

$$s_k \sim \mathcal{O}(k^{-1}), \tag{2.181}$$

and, in the general case, a series is said to possess algebraic convergence if, in the limit $k \to \infty$,

$$a_k \sim \mathcal{O}(k^{-q}), \tag{2.182}$$

where the constant, q, is the algebraic index of convergence. The rate of decrease of the coefficients, a_k, in such a case may be estimated by taking the ratio of two successive coefficients and noting (2.26)

$$\lim_{k \to \infty} \frac{a_{k+1}}{a_k} \sim \left[1 + \frac{1}{k}\right]^{-q},$$

$$\sim 1 - \frac{q}{k} + \frac{q(q+1)}{2k^2} \cdots, \qquad (2.183)$$

$$\sim 1. \qquad (2.184)$$

Because the asymptotic value in (2.183) is of the order of 1, series with an algebraic convergence rate could be slowly convergent and therefore series that converge faster than an algebraic convergent one are optimal. To establish other types of convergence consider the identity

$$\lim_{k \to \infty} k^q e^{-sk^r} = 0, \; r > 0, \qquad (2.185)$$

for constants q, s, r, from which it follows that a series with coefficients a_k, having a rate of decrease faster than k^{-q}, for all finite values of q, has exponential convergence. Therefore, for such a series, in the limit $k \to \infty$,

$$a_k \sim \mathcal{O}(e^{-sk^r}), \qquad (2.186)$$

with $r > 0$, and the exponential index of convergence, r, has a value obtained in the limit $k \to \infty$ from

$$r = \lim_{k \to \infty} \frac{ln|ln|a_k||}{ln(k)}. \qquad (2.187)$$

There is another possibility for the rate of decrease of the coefficients that combines both algebraic and exponential convergence characteristics and is called geometric convergence with the limiting form

$$a_k \sim \mathcal{O}(k^{-q} e^{-sk}). \qquad (2.188)$$

The expressions in (2.182)–(2.188) apply strictly for the case $k \to \infty$ and this limits their usefulness in practical applications (see also Sect. 2.3 of [47]). For this reason, an alternative method of measuring convergence behavior of the series (2.172) is proposed in Chap. 4. Nevertheless, from this brief discussion it may be concluded that the best series approximations are those that show exponential convergence and, as is seen in Chap. 4, this type of convergence is typical of Chebyshev series. However, the properties of the function being approximated can substantially impact convergence behavior. For example, approximation in the neighborhood of a singularity, or discontinuity, will show poor convergence properties. This is because it is impossible to approximate a singularity with polynomials of finite order. A classic example is polynomial approximation of a step function that leads to the so-called Gibbs phenomenon [28,48].

EXAMPLES 2.30–2.35

2.11 Differentiation and Integration of Monomial and Polynomial Series

Conversion of monomial series such as those in Sect. 2.3 into polynomial series, as those in Sect. 2.4 (and vice versa) is not uncommon and some more detailed examples are introduced here for later reference. Table 2.4 lists some polynomial expansions for powers of the argument and this is extended to higher orders for $\frac{sin(x)}{x}$ by using the monomial series of (2.23) divided by the argument x truncated at the x^{12} term.

The expansion in Legendre polynomials is shown in EXAMPLE 2.36 where each monomial term is first expanded in polynomials and those of the same order are collected to define the coefficients of the Legendre series expansion. The error in the polynomial expansion is tested at two argument values with a partial increment in the polynomial order, and the corresponding error curve is also shown on the interval [0, 1].

For the corresponding Chebyshev polynomials EXAMPLE 2.37 shows the same monomial expansion of $\frac{sin(x)}{x}$ and also the derivative. Firstly, the monomial series of the function and the term-by-term derivative of the monomial series are both expanded in Chebyshev polynomials and coefficients are collected for the Chebyshev expansion of both the function and the derivative. A simpler method uses the obvious property that the derivative of each Chebyshev polynomial may again be expanded as a Chebyshev series and coefficients of the series for the derivative may be obtained from those of the function itself if they are known. The coefficients of the Chebyshev series for the derivative may be derived from them by a simple recurrence as follows.

Given the truncated series

$$f(x) = \sum_{k=0}^{N\prime} h_k T_k(x), \qquad (2.189)$$

and the corresponding expansion of the derivative

$$\frac{df(x)}{dx} = \sum_{k=0}^{N\prime} h'_k T_k(x), \qquad (2.190)$$

then it may be shown that

$$h'_k = (2k+1)h_{k+1} + h'_{k+2}, \ k = N-1, \ldots, 0, \qquad (2.191)$$

where the downward recurrence is started with

$$\begin{aligned} h'_{N+1} &= 0 \\ h'_N &= 0. \end{aligned} \qquad (2.192)$$

These results are a special case of formulas for Chebyshev series derived in Chap. 3 of [4] using the property

$$\int T_k(x)dx = \frac{1}{2}\left\{\frac{T_{k+1}(x)}{k+1} - \frac{T_{k-1}(x)}{k-1}\right\}, \quad (2.193)$$

for a function where the series coefficients are known

$$g(x) = \sum_{k=0}^{N}{}' h'_k T_k(x), \quad (2.194)$$

the Chebyshev expansion coefficients of the integral

$$\int_{-1}^{x} g(x)dx = \sum_{k=0}^{N+1}{}' h_k T_k(x)dx, \quad (2.195)$$

are obtainable from the simple recurrence

$$h_k = \frac{h'_{k-1} - h'_{k+1}}{2k}, \quad k = 1, 2, \ldots, N - 1. \quad (2.196)$$

with starting values

$$h_{N+1} = \frac{h'_N}{2(N+1)},$$

$$h_N = \frac{h'_N}{2(N)}, \quad (2.197)$$

with the first coefficient obtained from the value of the series at $x = -1$

$$\frac{1}{2}h_0 = \sum_{k=1}^{N+1}(-1)^{k-1}h_k. \quad (2.198)$$

These results for the integral are seen to be the converse of the above results for the derivative.

In the case that the series are for the shifted Chebyshev case with $x \in [0, +1]$, the factor of $\frac{1}{2}$ in (2.193), (2.196), (2.197) is replaced by $\frac{1}{4}$. The Tables 2.2 and 2.4 are extended to the shifted Chebyshev polynomials in the Excel spreadsheet file.

EXAMPLES 2.36–2.39
SPREADSHEET XNTM.XLSX

2.12 Exercises

1. Consider the mappings of x into $t \in [-1, +1]$ used in EXAMPLE 2.1. Apply these to the function e^{-x^2}, with $x \in [0, +5]$ and $x \in [0, +\infty]$ to tabulate the respective values of x and the function value at $t = -0.5, -0.25, +0.25,$ and $+0.5$.

2. Plot the error curve $E(x)$ on $x \in [-\pi, +\pi]$ for the approximation $sin(x) \approx x - \frac{1}{3!}x^3 + \frac{1}{5!}x^5$ to $sin(x)$. From inspection, or otherwise, determine at which approximate values of x in the given range the error has a maximal value. Is the bound on the error given in Table 2.1 larger than this maximum value?

3. Use the special form (2.32) of the Taylor series for $f(x+h)$ in powers of h to obtain the 4th-order approximation to the polynomial $1 + x + x^2 + x^3 + x^4 + x^5$ on the interval $[-1, +1]$ by evaluating the derivatives at $x = 0$. Plot the error curve and estimate a bound on the error using the error term.

4. Tabulate the zeros of the Chebyshev polynomials $T_n(x)$ for $n = 2, 3, 4, 5, 6$ from (2.38). Tabulate also the distance between subsequent zeros for fixed n and plot the value of the zero closest to $+1$ for each n.

5. Using the expression for x^3 from Table 2.4 and the convolution results for Chebyshev series in Sect. 2.9, obtain the Chebyshev series coefficients of x^6.

6. Obtain the explicit expressions for the Legendre and Chebyshev polynomials, respectively, for

   (a) $P_2(x)$ and $P_3(x)$ using $P_0(x)$ and $P_1(x)$ in the recurrence (2.43),
   (b) $T_2(x)$ and $T_3(x)$ using $T_0(x)$ and $T_1(x)$ in the recurrence (2.45).

7. Obtain the explicit expressions for the Legendre polynomial $P_6(x)$ and the derivative $\frac{dP_6(x)}{dx}$ by inserting the expressions from Table 2.2 for $P_5(x)$ and $P_4(x)$ into

   (a) the recurrence (2.43) and
   (b) the recurrence (2.44).

8. Using the coefficients given in Table 2.4

   (a) apply the recurrence (2.67) to evaluate the Chebyshev expansion of (2.66) for x^4 at $x = 1$, and note that the first term in the series is $\frac{1}{2}h_0$,
   (b) apply the recurrence (2.67) to evaluate the Chebyshev expansion of (2.69) for x^5 at $x = 1$,
   (c) apply the recurrence (2.64) to evaluate the Chebyshev expansion of (2.63) for $x^2 + x^3$ at $x = 1$, and note that the first term in the series is $\frac{1}{2}h_0$.

2.13 Programming Problems

1. Using the algorithm in EXAMPLES 2.16–2.18 write a program in the coding language of your choice (Fortran or C) for the Fourier expansion of e^x and repeat the analysis up to order $K = 10$. Plot the error curve on the argument range $[-1, +1]$ and confirm results in EXAMPLES 2.16–2.18. Then raise the order to $K = 100$ and investigate the improvement in the error curve.
2. Write a program in the coding language of your choice (Fortran or C) for the Chebyshev expansion of the spherical Bessel function $j_\ell(z)$ using the algorithm in EXAMPLES 2.21–2.23 and repeat the analysis up to order $K = 10$, for $\ell = 0, 1, 2$. Plot the respective error curves on the argument range $[0, +1]$ and confirm the results in EXAMPLES 2.21–2.23. For which value of the argument is the error at a maximum value?
3. Write a program to compute the Chebyshev expansion coefficients of $sin(x)$ starting from the monomial series expansion similar to the algorithm in EXAMPLE 2.39 and compare the results to those in that example. Apply your results to compute the Chebyshev expansion coefficients of the derivative $\frac{d(sin(x))}{dx}$ using the same method of EXAMPLE 2.39 and tabulate the error curve against the exact result.

References

1. Wallace PR (1984) Mathematical analysis of physical problems. Dover Publications, New York
2. Watson GN (1966) A treatise on the theory of Bessel functions. Cambridge University Press, Cambridge, UK
3. Thomas GB (1960) Calculus and analytical geometry. Addison-Wesley Publishing Company Inc, Reading, MA
4. Fox L, Parker IB (1968) Chebyshev polynomials in numerical analysis. Oxford University Press, Oxford, UK
5. Rivlin TJ (1974) The Chebyshev polynomials. John Wiley and Sons, New York
6. Abramowitz M, Stegun IA (eds) (1970) Handbook of mathematical functions. Dover Publications, New York
7. Olver FWJ, Lozier DW, Boisvert RF, Clark CW (eds) (2010) NIST handbook of mathematical functions. NIST and Cambridge University Press, UK
8. Jeffreys H, Swirles B (Lady Jeffreys) (1962) Methods of mathematical physics. Cambridge University Press, Cambridge, UK
9. Tchebyshef PI (1962) Oeuvres. Chelsea, New York
10. Rivlin TJ (2020) The Chebyshev polynomials, 2nd edn. Dover Publications, New York, NY
11. Spanier J, Oldham KB (1987) An atlas of functions. Hemisphere Publishing Corporation, New York
12. Powell MJD (1981) Approximation theory and methods. Cambridge University Press, Cambridge, UK
13. Luke YL (1975) Mathematical functions and their approximations. Academic Press, New York
14. Luke YL (1969) The special functions and their approximation, vol I and II. Academic Press, New York

References

15. Tolstov GP (1976) Fourier series. Dover Publications, New York, NY
16. Clenshaw CW (1955) A note on the summation of chebyshev series. Math Comput 9:118–120
17. Clenshaw CW (1962) Mathematical tables, vol. 5. National Physical Laboratory. H.M. Stationary Office, London
18. Clenshaw CW (1957) The numerical solution of linear differential equations in Chebyshev series. Proc Camb Phil Soc 53:134–149
19. Clenshaw CW, Curtis AR (1960) A method for numerical integration on an automatic computer. Numer Math 2:197–205
20. Press WH, Teukolsky SA, Vetterling WT, Flannery BP (1992) Numerical recipes in FORTRAN, 2nd edn. Cambridge University Press
21. Luke YL (1977) Algorithms for the computation of mathematical functions. Academic Press, New York
22. Herivel J (1975) Joseph Fourier the man and the physicist. Oxford University Press, Oxford, UK
23. Van Loan C (1992) Computational frameworks for the fast Fourier transform. Society for Industrial and Applied Mathematics, Philadelphia, PA
24. Briggs WL, Henson VE (1995) The DFT an owner's manual for the discrete Fourier transform. Society for Industrial and Applied Mathematics, Philadelphia, PA
25. Brigham EO (1974) The fast Fourier transform. Prentice-Hall, Inc., Englewood Cliffs, NJ
26. Cooley JW, Tukey JW (1965) An algorithm for the machine calculation of complex Fourier series. Math Comput 19:297–301
27. Elliot DF, Ramamohan Rao K (1982) Fast transforms algorithms, analyses, applications. Academic Press, New York, NY
28. Carslaw HS (1950) An introduction to the theory of Fourier's series and integrals, 3rd edn. Dover Publications, New York, NY
29. Nussbaumer HJ (1982) Fast Fourier transform and convolution algorithms, 2nd edn. Springer-Verlag, Berlin
30. Oberhettinger F (1973) Fourier expansions a collection of formulas. Academic Press, New York, NY
31. Sneddon IN (1961) Special functions of mathematical physics and chemistry. Oliver and Boyd, Edinburgh, UK
32. Delic G (1979) Chebyshev series for the spherical Bessel function $JL(r)$. Comput Phys Commun 18:73–86
33. Delic G (1974) Optical model parameter searches for 16o+11b elastic scattering. Phys Lett 49B:412–414
34. Delic G (1976) Exact finite range dwba calculations for 13c(3he,6he)10c. Phys Rev C 14:619–624
35. Delic G (1975) Optical model analysis of n+c and c+c elastic scattering. Phys Rev Lett 34:1468–1471
36. Delic G (1976) Computing the finite range dwba for 11b(16o,15n)12c. Phys Rev Lett 36:569–572
37. Delic G (1979) Chebyshev expansion of the associated legendre polynomial $plm(x)$. Comput Phys Commun 18:63–71
38. Bateman H (1953) Higher transcendental functions, vol I. Bateman Manuscript Project. McGraw Hill, New York
39. Szegö G (1939) Orthogonal Polynomials, vol 23. Colloqium Publications. American Mathematical Society, New York
40. Delic G, Lemmer RH (1982) Quadrupole excitations in the two-center shell model (ii). Nucl Phys A 380:270–284
41. Bateman H (1953) Higher transcendental functions, vol II. Bateman Manuscript Project. McGraw Hill, New York

42. Whittaker ET, Watson GN (1965) A course of modern analysis. Cambridge University Press, Cambridge, UK
43. Magnus W, Oberhettinger F, Soni RP (1966) Formulas and theorems for the special functions of mathematical physics, 3rd edn. Springer-Verlag, Berlin
44. Delic G, Lemmer. RH (1985) Chebyshev series for the parabolic cylinder function dl(z) and derivatives with respect to order and argument. NTRG 85-7, Nuclear Theory Research Group, University of the Witwatersrand, Johannesburg
45. Delic G, Malherbe SM (1988) Subroutines for convolution sums of Fourier and Chebyshev series. Comput Phys Commun 48:305–312
46. Faires JD, Faires BT (1983) Calculus and analytical geometry. Prindle, Weber and Schmidt, Boston, MA
47. Boyd JP (2001) Chebyshev and Fourier spectral methods, 2nd edn. Dover Publications, New York, NY
48. Gottlieb D, Orszag SA (1977) Numerical analysis of spectral methods: theory and applications. Society for Industrial and Applied Mathematics, Philadelphia, PA

Interpolation of Discrete Data 3

3.1 Definition of the Interpolating Polynomial

If a function $f(x)$ is a continuous function in the argument $x \in [a,b]$ over the whole interval, then it has a representation in the form of (2.13), with a basis $\{\psi_K\}_1^\infty$ consisting of continuous smooth functions defined on the same interval. However, even if $f(x)$ is not continuous over the whole interval and is not known for some values of the argument x, then it may still be represented in the form of (2.13), but only with a finite set of basis elements $\{\psi_K\}_1^N$ each of which has only a finite number of derivatives.

For the purposes of this discussion, it is assumed that for each value of the argument x, there always is a unique value of the function $f(x)$ and that the case of two function values at the same argument does not occur. Such a situation would correspond to a step discontinuity and this would represent a function outside the class of continuous functions. Even so, the function itself, while it may be unknown for all x in the interval, may be known for a finite number of discrete argument values. The question then arises: how can the unknown function which generates these discrete values be determined? Such a function is, of course, continuous because it is defined for all values of $x \in [a,b]$ and it may, in general, have an infinite number of derivatives on this interval. But the construction of the unknown function would require an infinite number of points on the interval and only a finite number is available for this purpose. Therefore, it is not possible to exactly reconstruct the unknown function $f(x)$ because only a finite number of function values are available as data points.

Supplementary Information The online version contains supplementary material available at https://doi.org/10.1007/978-3-031-90178-2_3

However, it is possible to construct a function $p(x)$ which approximates $f(x)$ using the available data on $x \in [a, b]$.

Two situations are possible in constructing the function $p(x)$ which approximates $f(x)$. In the first case, the approximating function is chosen to reproduce exactly the available discrete data for the function. In the second case, the approximating function is required only to produce a "best fit" to the available discrete data without exactly reproducing each individual datum point. This second case is described in Chap. 4, after the concept of "best" fit has been clarified. The first case is discussed in this chapter because it is a simple concrete application of the discussion in Sect. 2.4. Here, just as in Chap. 2, a polynomial $p_{n-1}(x)$ of degree $n-1$ may be constructed to approximate $f(x)$ such that it reproduces the discrete data at n points. However, because the function $f(x)$ is not known, a different approach to that in Chap. 2 must be followed in choosing the basis elements $\{\psi_K\}_1^n$ and also for solution of the unknown coefficients $\{c_K\}_1^n$ in the linear expression

$$p_{n-1}(x) = \sum_{K=1}^{n} c_K \psi_K(x). \tag{3.1}$$

A general existence theorem for polynomial interpolation ensures that a polynomial $p_{n-1}(x)$, chosen such that it reproduces the discrete data exactly, always exists and is unique. Given a set $\{X_i\}_1^n$ of discrete data pairs $X_i = (x_i, f_i), i = 1, \ldots, n$, such that $x_i \neq x_j$ for $i \neq j$, then $p_{n-1}(x)$ is an interpolating polynomial on $x \in [a = x_1, b = x_n]$ if $p_{n-1}(x_i) = f_i, i = 1, \ldots, n$. Note that, if data pairs (x_i, f_i) are added to or subtracted from the data set $\{X_i\}_1^n$ such that $i = 1, \ldots, m$, and $n \neq m$, then another interpolating polynomial $p_{m-1}(x)$ is constructed such that $p_{n-1}(x) \neq p_{m-1}(x)$.

If $\{\psi_K\}_1^n$ is a set of polynomials, then (3.1) solves the interpolation problem by using them to construct an interpolating polynomial $p_{n-1}(x)$. In the following discussion, two methods to construct such interpolating polynomials are described. These methods are for the Newton[1] and Lagrange[2] form of the interpolating polynomial which differ only in the choice of basis element functions $\{\psi_K\}_1^n$ and the technique of solution for the corresponding unknown set of coefficients $\{c_K\}_1^n$. However, because both methods construct polynomials, then either method may be cast in the equivalent monomial form of (2.18) and therefore the interpolating polynomial of a fixed degree is unique irrespective of the method of its construction. Note that neither interpolation method requires the argument values to be equidistant. The next few sections discuss this general case and later sections develop (and tabulate) formulas useful in practical applications for the equidistant case. Some background material may be found in numerous sources including [1,2].

EXAMPLE 3.1

---

[1] Isaac Newton, English physicist and mathematician 1642–1727.
[2] Joseph Louis Lagrange, Italian-French mathematician and theoretical physicist 1736–1813.

3.2 The Newton Interpolating Polynomial

The Newton form of the interpolating polynomial is attractive because it augments by one unit the degree of an existing interpolating polynomial in using an additional datum pair (x_i, f_i). This method then solves for an additional coefficient in the set $\{c_K\}_1^n$ without the need of recalculating the preceding ones. The procedure is demonstrated for the set of data pairs (x_i, f_i), $i = 1, \ldots, n$, for which the lowest order interpolating polynomial has degree zero

$$p_0(x) = f_1, \tag{3.2}$$

and is the straight line parallel to the abscissa passing through f_1. The next highest order uses two data points to construct a straight line passing through both data values and does this by augmenting $p_0(x)$

$$p_1(x) = p_0(x) + c_2(x - x_1), \tag{3.3}$$

where, in this indexing notation, the first coefficient $c_1 = f_1$. Then the second coefficient c_2 is solved for by using the second data point, where direct substitution into (3.3) gives

$$f_2 = f_1 + c_2(x - x_1), \tag{3.4}$$

and the constant c_2 is the slope of a straight line passing through the data pairs $(x_1, f_1), (x_2, f_2)$. Similarly, a third datum pair (x_3, f_3) may be used to solve for the new coefficient c_3 in the augmented polynomial

$$p_2(x) = p_1(x) + c_3(x - x_1)(x - x_2). \tag{3.5}$$

As a demonstration, consider construction of the Newton interpolating polynomial for the three data pairs $(0, 3)$, $(-1, 5)$ and $(1, -5)$. Here, $p_0(x) = 3$ and solving for c_2 in (3.3)

$$5 = 3 + c_2(-1 - 0), \tag{3.6}$$

giving $c_2 = -2$ and therefore, substitution into (3.3) gives the result

$$p_1(x) = p_0(x) - 2x. \tag{3.7}$$

Finally, the last coefficient in (3.5) is solved using the last datum pair

$$-5 = 1 + c_3(1 - 0)[1 - (-1)], \tag{3.8}$$

and therefore

$$p_2(x) = p_1(x) - 3x(x + 1). \tag{3.9}$$

In numerical applications which require approximate function values at $x \neq x_i$ it is best not to write out the corresponding monomial form for each order. This would require redundant work and it is more efficient to compute lowest to highest order Newton polynomials successively. To demonstrate this technique, consider the above example computed for the value $x = \frac{1}{2}$

$$p_0\left(\frac{1}{2}\right) = 3,$$

$$p_1\left(\frac{1}{2}\right) = p_0\left(\frac{1}{2}\right) - 2\left(\frac{1}{2}\right),$$
$$= 3 - 1,$$
$$= 2,$$

$$p_2\left(\frac{1}{2}\right) = p_1\left(\frac{1}{2}\right) - 3\left(\frac{1}{2}\right)\left(\frac{1}{2} + 1\right),$$
$$= 2 - 3 \times \frac{3}{4},$$
$$= -\frac{1}{4}. \tag{3.10}$$

From the foregoing discussion, the general form of the Newton interpolating polynomial of degree k is seen to have the form

$$p_k(x) = p_{k-1}(x) + c_{k+1}(x - x_1)(x - x_2) \cdots (x - x_k), \tag{3.11}$$

or, explicitly

$$p_k(x) = c_1 + \sum_{i=2}^{k+1} c_i \left[\prod_{j=1}^{i-1}(x - x_j) \right], \tag{3.12}$$

where

$$\prod_{j=1}^{k}(x - x_j) = (x - x_1)(x - x_2) \cdots (x - x_k), \tag{3.13}$$

with the definition

$$\prod_{j=1}^{0}(x - x_j) = 1. \tag{3.14}$$

The polynomial (3.12) interpolates $k + 1$ data points $\{X_i\}_1^{k+1}$ when the last point X_{k+1} is substituted into (3.12) to solve for the last coefficient c_{k+1}. Note that the degree of the interpolating Newton polynomial is always one less than the total number of points in the set $\{X_i\}_1^{k+1}$.

EXAMPLES 3.2 and 3.3

3.3 Divided Differences and Newton Interpolation

Consider again the expressions of (3.2)–(3.5) used to solve for the set of constants $\{c_K\}_1^3$ written in the form

$$c_1 = f_1,$$

$$c_2 = \frac{f_2 - f_1}{x_2 - x_1},$$

$$c_3 = \frac{1}{x_3 - x_2}\left(\frac{f_3 - f_1}{x_3 - x_1} - c_2\right), \tag{3.15}$$

$$= \frac{1}{x_3 - x_1}\left(\frac{f_3 - f_2}{x_3 - x_2} - \frac{f_2 - f_1}{x_2 - x_1}\right), \tag{3.16}$$

and the last result follows from some lengthy algebra, that may be confirmed by showing the difference of (3.15) and (3.16) is zero. The result of (3.16) has an interesting structure in that on defining

$$f(x_i) = f_i \tag{3.17}$$

and

$$f[x_i, x_{i+1}] = \frac{f[x_{i+1}] - f[x_i]}{x_{i+1} - x_i}, \tag{3.18}$$

then (3.16) takes the form

$$c_3 = \frac{f[x_2, x_3] - f[x_1, x_2]}{x_3 - x_1}, \tag{3.19}$$

where, in this last expression, the divisor is the difference of alternate abscissas in the data pairs and not subsequent ones as in (3.18). The result (3.19) may also be generalized as

$$f[x_i, x_{i+1}, x_{i+2}] = \frac{f[x_{i+1}, x_{i+2}] - f[x_i, x_{i+1}]}{x_{i+2} - x_i} \tag{3.20}$$

and the quantity denoted as $f[x_i, x_{i+1}]$ is called the first divided difference at x_i while $f[x_i, x_{i+1}, x_{i+2}]$ is the second divided difference at x_i and $f[x_i]$ denotes the divided difference of zero order. In general, the $k - 1$-order divided difference is defined in terms of the recursive property

$$f[x_1, x_2, \ldots, x_k] = \frac{f[x_2, x_3, \ldots, x_k] - f[x_1, x_2, \ldots, x_{k-1}]}{x_k - x_1}, \tag{3.21}$$

and this recurrence is applicable to divided differences of any order

$$f[x_i, x_{i+1}, \ldots, x_k] = \frac{f[x_{i+1}, x_{i+2}, \ldots, x_k] - f[x_i, x_{i+1}, \ldots, x_{k-1}]}{x_k - x_i}. \tag{3.22}$$

Table 3.1 Divided differences

Argument	Zeroth order	First order	Second order
x_1	$f[x_1]$		
		$f[x_1, x_2]$	
x_2	$f[x_2]$		$f[x_1, x_2, x_3]$
		$f[x_2, x_3]$	
x_3	$f[x_3]$		

Another important property of divided differences is the invariance property which states that the divided difference $f[x_i, x_{i+1}, \ldots, x_k]$ is invariant under all permutations of the arguments $x_i, x_{i+1}, \ldots, x_k$. This is most easily understood for the second divided difference $f[x_1, x_2, x_3] = f[x_3, x_1, x_2]$ because this divided difference is the coefficient c_3 in (3.19) and this is the coefficient of the quadratic power of x in the interpolating polynomial $p_2(x)$ of (3.5). This quadratic interpolating polynomial is the same irrespective of the order in which the data pairs are used and therefore the value of the coefficient c_3 is unique.

Other expressions for the second divided difference would follow from explicit solution in (3.16), whence

$$f[x_1, x_2, x_3] = \frac{f_3 - f[x_1] - f[x_1, x_2](x_3 - x_1)}{(x_3 - x_1)(x_3 - x_2)} \tag{3.23}$$

and the generalization is

$$f[x_1, x_2, \ldots, x_k] = \frac{f_k - \sum_{i=1}^{k-1} f[x_1, x_2, \ldots, x_i] \prod_{j=1}^{i-1}(x_k - x_j)}{\prod_{j=1}^{k-1}(x_k - x_j)}, \tag{3.24}$$

which is a considerably more complicated expression than (3.21) for the $(k-1)$th-order divided difference. Furthermore, (3.24) requires redundant computations and therefore the recurrence relation (3.22) is usually applied in the form of a divided difference table with a simple example shown in Table 3.1.

Note that the divided difference table is triangular and that there are as many columns in the table as there are data points (if the zeroth-order column is included). What is important about Table 3.1 is that the coefficients, c_1, c_2 and c_3, in the Newton form of the interpolating polynomial (3.12), are the entries in the uppermost diagonal. The divided difference table is the simplest way to compute the coefficients of the Newton interpolating polynomial. Consider again the case discussed in Sect. 3.2 for which the completed divided difference table is shown in Table 3.2.

3.4 The Lagrange Interpolating Polynomial

Table 3.2 Divided difference values

Argument	Zero order	First order	Second order
0	3		
		−2	
−1	5		−3
		−5	
1	−5		

EXAMPLES 3.4 and 3.5

3.4 The Lagrange Interpolating Polynomial

In the case of two data points $\{X_i\}_1^2$, the highest order interpolating polynomial is $p_1(x)$ which is a straight line. One method of constructing $p_1(x)$ is to define two polynomials

$$\ell_{11}(x) = \frac{x - x_2}{x_1 - x_2}, \tag{3.25}$$

$$\ell_{12}(x) = \frac{x - x_1}{x_2 - x_1}, \tag{3.26}$$

that have the property

$$\ell_{11}(x_1) = \ell_{12}(x_2) = 1, \tag{3.27}$$

and

$$\ell_{11}(x_2) = \ell_{12}(x_1) = 0. \tag{3.28}$$

Obviously, the divisors in (3.25), (3.26) normalize the polynomials $\ell_{11}(x)$, $\ell_{12}(x)$ to unity in (3.27). The Lagrange interpolating polynomial is then defined by the linear combination

$$p_1(x) = f_1 \ell_{11}(x) + f_2 \ell_{12}(x). \tag{3.29}$$

Note that the reason for the normalization (3.27) becomes evident, when $x = x_1, x_2$ is substituted into (3.29) because it follows from (3.25), (3.26) that $p_1(x_1) = f_1$ or $p_1(x_2) = f_2$, respectively. If one more datum pair is added, then there are three data points $\{X_i\}_1^3$, and three polynomials are defined as

$$\ell_{21}(x) = \ell_{11}(x) \frac{(x - x_3)}{(x_1 - x_3)},$$

$$\ell_{22}(x) = \ell_{12}(x) \frac{(x - x_3)}{(x_2 - x_3)},$$

$$\ell_{23}(x) = \frac{(x - x_1)}{(x_3 - x_1)} \frac{(x - x_2)}{(x_3 - x_2)}. \tag{3.30}$$

The Lagrange interpolating polynomial in this case is the quadratic polynomial defined by the linear combination

$$p_2(x) = f_1 \ell_{21}(x) + f_2 \ell_{22}(x) + f_3 \ell_{23}(x), \tag{3.31}$$

with the discrete orthogonality property

$$\ell_{2i}(x_j) = 1, i = j, \tag{3.32}$$
$$\ell_{2i}(x_j) = 0, i \neq j. \tag{3.33}$$

As the number of data points increases, the order of the interpolating polynomial may be increased, but at the cost of computing a new set of polynomial basis elements $\{\ell_{k-1\,i}\}_1^k$ for k data points. These are constructed from the general expression

$$\ell_{k-1\,i}(x) = \frac{(x - x_1)}{(x_i - x_1)} \frac{(x - x_2)}{(x_i - x_2)} \cdots \frac{(x - x_{i-1})}{(x_i - x_{i-1})} \frac{(x - x_{i+1})}{(x_i - x_{i+1})} \cdots \frac{(x - x_k)}{(x_i - x_k)}, \tag{3.34}$$

where the term in x_i is missing because the divisor is zero. These polynomials satisfy the orthogonality relations (3.32), (3.33) and the general Lagrange interpolating polynomial of degree $k - 1$ is then given by the linear combination

$$p_{k-1}(x) = \sum_{i=1}^{k} f_i \ell_{k-1\,i}(x) \tag{3.35}$$

of which (3.31) is a special case with $k = 3$.

The disadvantage of the Lagrange form of the interpolating polynomial is that a different set of basis polynomials, $\{\ell_{k-1\,i}\}_1^k$, must be calculated if data points are added, or the degree of the interpolating polynomial is increased. On the other hand, the Newton form of the interpolating polynomial requires computation of the divided difference table and this becomes cumbersome if the order of the interpolating polynomial is large. Note that neither interpolation method requires the argument values to be equidistant.

EXAMPLES 3.6–3.8

3.5 Error in Polynomial Interpolation

If a polynomial of degree $k-1$ interpolates a set of discrete data $X_i = (x_i, f_i)$, $i = 1, \ldots, k$, and $p_{k-1}(x_i) = f_i$, $i = 1, \ldots, k$, then it is meaningful to ask how does the error function

$$E(x) = f(x) - p_{k-1}(x), \tag{3.36}$$

behave on $x \in [a, b]$? In particular, does it follow that, for all x, $|E(x)| \to 0$, when $k \to \infty$? Some theoretical aspects of these questions are the subject of discussion in Chap. 4 and in this section known results for polynomial interpolates are summarized.

Two important results for divided differences are the following. If the function $f(x)$ is a polynomial of degree k, then all divided differences $f[x_1, x_2, \ldots, x_i]$, are zero when $i > k+1$. Furthermore, the kth-order divided difference is simply related to the kth derivative of a function $f(x)$ through

$$f[x_1, x_2, \ldots, x_{k+1}] = \frac{1}{k!} f^{(k)}(\xi) \tag{3.37}$$

for $x_1 \leq \xi \leq x_{k+1}$. This result is applied in a statement for the form of the interpolation error.

Let $p_{k-1}(x)$ interpolate $f(x)$ at k data points $\{X_i\}_1^k$. Then for any x, $x_1 \leq x \leq x_k$, an ξ exists, for continuous $f^{(k)}(x)$, such that the error function for the interpolation is given by either one of two equivalent expressions

$$f(x) - p_{k-1}(x) = \frac{1}{k!} f^{(k)}(\xi) \prod_{j=1}^{k} (x - x_j), \tag{3.38}$$

or, from (3.37),

$$f(x) - p_{k-1}(x) \approx f[x_1, x_2, \ldots, x_k, x] \prod_{j=1}^{k} (x - x_j). \tag{3.39}$$

Both (3.38) and (3.39) imply the use of $f(x)$ which is unknown and therefore these expressions for the error cannot be applied as they stand. However, if one more datum pair (x_{k+1}, f_{k+1}), is added, then an estimate of the error, (3.39), is

$$f(x) - p_{k-1}(x) \approx f[x_1, x_2, \ldots, x_k, x_{k+1}] \prod_{j=1}^{k} (x - x_j), \tag{3.40}$$

which, from (3.11), is seen to be the term which augments the polynomial $p_{k-1}(x)$ to give the next highest order Newton polynomial. Therefore, in place of (3.36), the estimate is

$$E(x) \approx p_k(x) - p_{k-1}(x), \tag{3.41}$$

because $f(x) \approx p_k(x)$ and this approximation is reliable as long as the sequence $\{p_i(x)\}_1^\infty$ is convergent. However, this is not always the case and caution is appropriate in application of (3.41).

EXAMPLE 3.9

3.6 The Derivative of the Interpolating Polynomial

Often not only the function value needs to be computed at non-tabular values of argument but also the derivatives of the function. One method of estimating the derivative is to approximate it by the derivative of the interpolating polynomial. To demonstrate this method, consider the lowest order Newton interpolating polynomial

$$p_1(x) = p_0(x) + f[x_1, x_2](x - x_1), \tag{3.42}$$

and the derivative at x for $x_1 \leq x \leq x_2$, is

$$p_1'(x) = f[x_1, x_2], \tag{3.43}$$

which is identical to (3.37) with the prime denoting the first derivative with respect to the argument. If it is assumed that the abscissas in the data are regularly spaced, then the two expressions for the derivative at x are obtainable from (3.43) depending as

(i) $x_1 = x, x_2 = x + h$, when

$$f'(x) \approx \frac{f(x+h) - f(x)}{h}, \tag{3.44}$$

(ii) $x_1 = x - h, x_2 = x + h$, when a symmetric form is

$$f'(x) \approx \frac{f(x+h) - f(x-h)}{2h}. \tag{3.45}$$

A higher degree approximation to the derivative of $f(x)$ follows from the derivative of the next highest order interpolating polynomial,

$$p_2(x) = p_0(x) + f[x_1, x_2](x - x_1) + f[x_1, x_2, x_3](x - x_1)(x - x_2), \tag{3.46}$$

for which the derivative expression is

$$p_2'(x) = f[x_1, x_2] + f[x_1, x_2, x_3](2x - x_1 - x_2). \tag{3.47}$$

The second term on the right hand side is seen to be a higher order correction to the first-order estimate of the derivative in (3.43). This correction is zero for the symmetric choice of case (ii) which means that (3.45) is a more accurate estimate for the first derivative than (3.44).

The next higher degree approximation to the derivative of $f(x)$ follows from the derivative of the third degree interpolating polynomial

$$p_3(x) = p_0(x) + f[x_1, x_2](x - x_1) + f[x_1, x_2, x_3](x - x_1)(x - x_2) \\ + f[x_1, x_2, x_3, x_4](x - x_1)(x - x_2)(x - x_3), \quad (3.48)$$

which gives the expression

$$p_3'(x) = f[x_1, x_2] + f[x_1, x_2, x_3](2x - x_1 - x_2) + f[x_1, x_2, x_3, x_4] \\ \times [(x - x_2)(x - x_3) + (x - x_1)(x - x_3) + (x - x_1)(x - x_2)]. \quad (3.49)$$

When the symmetric case of (ii) is extended such that $x_3 = x - 2h$, $x_4 = x + 2h$, the approximation to the derivative is

$$f'(x) \approx \frac{f(x+h) - f(x-h)}{2h} \\ + \frac{1}{12h} \{2[f(x+h) - f(x-h)] - [f(x+2h) - f(x-2h)]\}, \quad (3.50)$$

and the second term is a correction to the lower order estimate of the derivative (3.45). Higher order derivatives may similarly be obtained by continued differentiation of an interpolating polynomial of suitable degree. However, the simple method discussed in the next section accelerates this convergence.

EXAMPLE 3.10

3.7 Accelerating Convergence for Derivatives

The discussion in the previous section suggests that more accurate approximations to the derivative may only be obtained by successively increasing the order of the polynomial interpolation. However, as a general demonstration of alternative methods for enhancing convergence, consider the Richardson extrapolation technique. From Taylor's theorem (2.32), (2.33)

$$f(x+h) = f(x) + hf^{(1)}(x) + \frac{1}{2}h^2 f^{(2)}(\xi) + \cdots, \quad (3.51)$$

from which it follows that

$$f^{(1)}(x) = \frac{f(x+h) - f(x)}{h} - \frac{1}{2}h f^{(2)}(\xi), \quad (3.52)$$

for $x \leq \xi \leq x+h$ and the error term in this last result is $\mathcal{O}(h)$ which would represent slow convergence with decreasing h. On the other hand, consider again the Taylor series (2.32) from which it follows that

$$f(x \pm h) = f(x) \pm hf^{(1)}(x) + \frac{1}{2}h^2 f^{(2)}(x) \pm \frac{1}{6}h^3 f^{(3)}(x) + \cdots, \qquad (3.53)$$

which may be used to obtain the following expression

$$f^{(1)}(x) = \frac{f(x+h) - f(x-h)}{2h} - \frac{1}{6}h^2 f^{(3)}(x) - \frac{1}{120}h^4 f^{(4)}(x) + \cdots, \quad (3.54)$$

which has an error term $\mathcal{O}(h^2)$. Comparison of the error terms in (3.52) and (3.54) shows why (3.45) is a better estimate than (3.44). For a fixed function $f(x)$ and argument x, (3.54) may be considered a function of h and written in the form

$$f^{(1)}(x) = \phi(h) + \alpha_2 h^2 + \alpha_4 h^4 + \cdots \qquad (3.55)$$

with constants $\alpha_2, \alpha_4, \ldots$, and the function $\phi(h)$ defined as

$$\phi(h) = \frac{f(x+h) - f(x-h)}{2h}. \qquad (3.56)$$

The form of the series in (3.55) suggests that the accuracy in approximating $f^{(1)}(x)$ by $\phi(h)$ may be increased by decreasing h. On the other hand, as shown in Sect. 3.6, another approach is to increase the order of the function used to compute the approximant. There is yet another possibility which consists of applying (3.55), (3.56) as follows:

$$\phi(h) = f'(x) - \alpha_2 h^2 - \alpha_4 h^4 + \cdots, \qquad (3.57)$$

and

$$\phi(h/2) = f'(x) - \alpha_2 (h/2)^2 - \alpha_4 (h/2)^4 + \cdots, \qquad (3.58)$$

both of which have an error term $\mathcal{O}(h^2)$. However, these last two results may be applied to produce the linear combination

$$\frac{1}{3}[4\phi(h/2) - \phi(h)] = f'(x) + \frac{1}{4}\alpha_4 h^4 + \cdots, \qquad (3.59)$$

which has an error term $\mathcal{O}(h^4)$ and therefore represents a significantly better approximant for the derivative. Note the choice of the linear combination in (3.59) which eliminates the leading term $\mathcal{O}(h^2)$ in the error. The same technique may be repeated in (3.59) to eliminate the leading term $\mathcal{O}(h^4)$ in the error. This method is called Richardson extrapolation and may be continued to any desired order in the error.

EXAMPLE 3.11

3.8 Finite Differences

Divided differences were introduced in Sect. 3.3 as a means of constructing the coefficients of an interpolating polynomial for data pairs (x_i, f_i) with unequal spacing of the abscissa. This section introduces finite differences [1,2], as a generalization, for data pairs (x_i, f_i) equally spaced in x_i such that for h a constant

$$x_i = x_0 + ih, i = \pm 1, \pm 2, \ldots, \pm n. \tag{3.60}$$

For convenience in the remainder of this chapter, the discrete index i is allowed to assume negative values. Then the first forward, backward, and central finite difference is defined, respectively, by the operators Δ, ∇, and δ as follows:

$$\begin{aligned} f_{i+1} - f_i &= \Delta f_i, \\ &= \nabla f_{i+1}, \\ &= \delta f_{i+\frac{1}{2}}. \end{aligned} \tag{3.61}$$

The nomenclature forward (Table 3.3), backward (Table 3.4) and central (Table 3.5) is associated with the slope of the line joining differences with the same suffix in the difference table (as indicated by arrows in the respective tables). Note that the numbers in the respective columns of differences are the same and what differs is the notation that expresses the value of x with which each difference is associated.

Higher order differences follow from repeated application of the definitions in (4.61), or inspection of Tables 3.3, 3.4, and 3.5, so that for the second order

$$\begin{aligned} \Delta^2 f_i &= \Delta f_{i+1} - \Delta f_i, \\ \nabla^2 f_i &= \nabla f_i - \nabla f_{i-1}, \\ \delta^2 f_i &= \delta f_{i+\frac{1}{2}} - \delta f_{i-\frac{1}{2}}. \end{aligned} \tag{3.62}$$

Table 3.3 Forward differences

Abscissa	Zero order	First order	Second order	Third order
x_{-2}	f_{-2}		$\Delta^2 f_{-3}$	
		Δf_{-2}		$\Delta^3 f_{-3}$
x_{-1}	f_{-1}		$\Delta^2 f_{-2}$	
		Δf_{-1}		$\Delta^3 f_{-2}$
x_0	f_0 ↘		$\Delta^2 f_{-1}$	
		Δf_0 ↘		$\Delta^3 f_{-1}$
x_1	f_1		$\Delta^2 f_0$ ↘	
		Δf_1		$\Delta^3 f_0$
x_2	f_2		$\Delta^2 f_1$	

Table 3.4 Backward differences

Abscissa	Zero order	First order	Second order	Third order
x_{-2}	f_{-2}		$\nabla^2 f_{-1}$	
		∇f_{-1}		$\nabla^3 f_0$
x_{-1}	f_{-1}		$\nabla^2 f_0$ ↗	
		∇f_0 ↗		$\nabla^3 f_1$
x_0	f_0 ↗		$\nabla^2 f_1$	
		∇f_1		$\nabla^3 f_2$
x_1	f_1		$\nabla^2 f_2$	
		∇f_2		$\nabla^3 f_3$
x_2	f_2		$\nabla^2 f_3$	

Table 3.5 Central differences

Abscissa	Zero order	First order	Second order	Third order	Fourth order
x_{-2}	f_{-2}		$\delta^2 f_{-2}$		$\delta^4 f_{-2}$
		$\delta f_{-\frac{3}{2}}$		$\delta^3 f_{-\frac{3}{2}}$	
x_{-1}	f_{-1}		$\delta^2 f_{-1}$		$\delta^4 f_{-1}$
		$\delta f_{-\frac{1}{2}}$		$\delta^3 f_{-\frac{1}{2}}$	
x_0	f_0 →		$\delta^2 f_0$ →		$\delta^4 f_0$
		$\delta f_{\frac{1}{2}}$		$\delta^3 f_{\frac{1}{2}}$	
x_1	f_1		$\delta^2 f_1$		$\delta^4 f_1$
		$\delta f_{\frac{3}{2}}$		$\delta^3 f_{\frac{3}{2}}$	
x_2	f_2		$\delta^2 f_2$		$\delta^4 f_2$

Expressed in terms of function values the second-order differences are given by

$$\Delta^2 f_i = f_{i+2} - 2f_{i+1} + f_i,$$
$$\nabla^2 f_i = f_i - 2f_{i-1} + f_{i-2},$$
$$\delta^2 f_i = f_{i+1} - 2f_i + f_{i-1}, \tag{3.63}$$

and it follows that

$$\Delta^2 f_{i-1} = \nabla^2 f_{i+1} = \delta^2 f_i, \tag{3.64}$$

which is obvious from Tables 3.3, 3.4, and 3.5, for $\delta^2 f_0$, as an example, because this value has the same location in all three cases.

Higher order finite differences are obtainable from lower order ones by a generalization of (3.62), as in the case of forward differences

$$\Delta^m f_i = \Delta^{m-1} f_{i+1} - \Delta^{m-1} f_i. \tag{3.65}$$

Repeated application of (3.61) and (3.62) gives an expression of any power of the forward finite difference in terms of tabular function values

$$\Delta^m f_0 = \sum_{j=0}^{m} (-1)^j \binom{m}{j} f_{m-j}, \tag{3.66}$$

where the binomial coefficient is defined as

$$\binom{m}{j} = \frac{m(m-1)\cdots(m-j+1)}{m!}. \tag{3.67}$$

Alternatively, it is possible to derive an expression for a tabular value in an expansion of powers of the forward difference [1]

$$f_m = \sum_{j=0}^{m} \binom{m}{j} \Delta^j f_0, \tag{3.68}$$

and similar results can be obtained for backward or central differences. These results are of interest, but what is needed is an analogue of (3.68) for the case that an interpolate is required at some value of argument between tabular points. This is the subject of the remainder of this chapter beginning with the next section.

3.9 The Newton Polynomial Revisited

Consider again the Newton form of the interpolating polynomial given by (3.2)–(3.5) and the expressions for the coefficients of the interpolating polynomial in (3.16). It follows from application of the definition (3.60) for the forward difference for equidistant data that

$$\begin{aligned} c_1 &= f_1, \\ c_2 &= \frac{1}{h} \Delta f_1, \\ c_3 &= \frac{1}{2h^2} \Delta^2 f_1, \end{aligned} \tag{3.69}$$

and like powers of the finite difference and the variable x are seen to occur together in (3.2)–(3.5). It is of some interest to seek a general expression for this result that is analogous to that for (3.68) and in the following we follow Sect. 1.3 of [2]. Consider the tabular data at argument values given by (3.60). For $n+1$ data points, $\{X_i\}_0^n$, an interpolatory polynomial of degree n may be constructed. Following Sect. 1.3 of [2], on introducing the change in variable

$$z = \frac{x - x_0}{h},$$
$$x = x_0 + zh, \tag{3.70}$$

comparison with (3.60) shows that $z = i$ at tabular values of the argument and we seek an interpolatory polynomial in argument z. By analogy with (3.67), define a factorial polynomial $z^{(j)}$ by

$$\frac{z^{(j)}}{j!} = \binom{z}{j}, \tag{3.71}$$

and by comparison with (3.67) it follows that

$$z^{(j)} = z(z-1)(z-2)\cdots(z-j+1), \quad j = 1, 2, \ldots \tag{3.72}$$

with the first few members of this series defined by

$$z^{(0)} = 1,$$
$$z^{(1)} = z,$$
$$z^{(2)} = z(z-1),$$
$$z^{(3)} = z(z-1)(z-2).$$

Application of the forward difference operator for an interval of 1 and use of the definition (3.61) shows that

$$\Delta z^{(m)} = (z+1)^{(m)} - z^{(m)},$$
$$= (z+1)z(z-1)(z-2)\cdots(z-m+2)$$
$$- z(z-1)(z-2)\cdots(z-m+1),$$
$$= z(z-1)(z-2)\cdots(z-m+2)[(z+1) - (z-m+1)],$$
$$= mz^{(m-1)}. \tag{3.73}$$

From this result and the definition (3.71), it also follows that

$$\Delta \binom{z}{m} = \frac{mz^{(m-1)}}{m!},$$
$$= \binom{z}{m-1}. \tag{3.74}$$

Using the factorial polynomials, construct an interpolating polynomial of degree n that passes through $n + 1$ tabular points

$$p(z) = b_0 + b_1 \binom{z}{1} + b_2 \binom{z}{2} + \cdots + b_j \binom{z}{j} + \cdots + b_n \binom{z}{n}, \tag{3.75}$$

3.9 The Newton Polynomial Revisited

and solve for the coefficients $\{b_j\}_0^n$ by repeated application of the difference operator Δ

$$\Delta p(z) = b_1 + b_2 \binom{z}{1} + b_3 \binom{z}{2} + \cdots + b_j \binom{z}{j-1} + \cdots + b_n \binom{z}{n-1}, \tag{3.76}$$

$$\Delta^2 p(z) = b_2 + b_3 \binom{z}{1} + b_4 \binom{z}{2} + \cdots + b_j \binom{z}{j-2} + \cdots + b_n \binom{z}{n-2}, \tag{3.77}$$

$$\Delta^m p(z) = b_m + b_{m+1} \binom{z}{1} + \cdots + b_n \binom{z}{n-m}, \tag{3.78}$$

where, for the first term, $\Delta\binom{z}{0} = 0$, from (3.74). Choosing $z = 0$ in each of (3.75) to (3.78) solves for the coefficients

$$\begin{aligned} b_0 &= p(0), \\ b_1 &= \Delta p(0), \\ b_m &= \Delta^m p(0), \end{aligned} \tag{3.79}$$

and because the first tabular value is $p(0) = f_0$, it follows that the interpolating polynomial (3.75) of degree n is

$$p_n(z) = \sum_{m=0}^{n} \binom{z}{m} \Delta^m f_0, \tag{3.80}$$

which is the analogue of (3.68). At non-tabular values of the argument, when $z \neq i$ in (3.60), $f(x)$ can be approximated by the interpolating polynomial in (3.80)

$$f_z \approx p_n(z) \tag{3.81}$$

where the subscript n denotes the rank of the approximating polynomial. This expression uses the forward differences of Table 3.3 and is therefore the Newton interpolatory polynomial. A similar method may be applied to find the corresponding expression for Newton backward differences

$$p_n(z) = \sum_{m=0}^{n} (-1)^m \binom{-z}{m} \nabla^m f_0 \tag{3.82}$$

which uses the backward differences of Table 3.4 (see Sect. 1.4 of [2]).

3.10 Stirling's Central Difference Formula

In the previous section, an interpolatory polynomial was developed using either forward or backward differences. In this section, a special form of the interpolatory polynomial is derived that uses a mean of the two, based on Stirling's central difference formula as developed in Sect. 1.5 of [2]. Special cases are tabulated in the next section for practical application in later chapters. The first step is to derive Gauss's forms of the interpolatory polynomial.

With tabular points at $z = i$, corresponding to $z = 0, 1, -1, 2, -2, \ldots$, in (3.60), define an interpolatory polynomial

$$p(z) = b_0 + b_1 \binom{z}{1} + b_2 \binom{z}{2} + b_3 \binom{z+1}{3} + b_4 \binom{z+1}{4} + \cdots + b_{2j} \binom{z+j-1}{2j} + b_{2j+1} \binom{z+j}{2j+1} + \cdots, \qquad (3.83)$$

as in the previous section, repeated application of the forward difference operator and the result (3.74), gives

$$\Delta p(z) = b_1 + b_2 \binom{z}{1} + b_3 \binom{z+1}{2} + b_4 \binom{z+1}{3} + \cdots + b_{2j} \binom{z+j-1}{2j-1} + b_{2j+1} \binom{z+j}{2j} + \cdots, \qquad (3.84)$$

$$\Delta^2 p(z) = b_2 + b_3 \binom{z+1}{2} + b_4 \binom{z+1}{2} + \cdots + b_{2j} \binom{z+j-1}{2j-2} + b_{2j+1} \binom{z+j}{2j-1} + \cdots, \qquad (3.85)$$

or, in the general case for even and odd powers of the forward difference

$$\Delta^{2j} p(z) = b_{2j} + b_{2j+1} \binom{z+j}{1} + \cdots, \qquad (3.86)$$

$$\Delta^{2j+1} p(z) = b_{2j+1} + b_{2j+2} \binom{z+j}{1} + \cdots. \qquad (3.87)$$

Placing $z = 0$ in (3.83) and $z = 0, -1, \ldots, -j, -j, \ldots$, in (3.84)–(3.87), solves for the coefficients

$$b_0 = p(0),$$
$$b_1 = \Delta p(0),$$
$$b_2 = \Delta^2 p(-1),$$
$$\cdots\cdots$$

3.10 Stirling's Central Difference Formula

$$b_{2j} = \Delta^{2j} p(-j),$$
$$b_{2j+1} = \Delta^{2j+1} p(-j), \tag{3.88}$$

and because the argument values in (3.88) are tabular values, it follows that using (3.83) leads to Gauss's forward difference approximation to $f(z)$, for $z \neq i$

$$f_z \approx f_0 + \binom{z}{1}\Delta f_0 + \binom{z}{2}\Delta^2 f_{-1} + \binom{z+1}{3}\Delta^3 f_{-1} + \cdots$$
$$+ \binom{z+j-1}{2j}\Delta^{2j} f_{-j} + \binom{z+j}{2j+1}\Delta^{2j+1} f_{-j} + \cdots. \tag{3.89}$$

This expression may be written in terms of central differences on noting the following equivalences from (3.61), $\delta f_{\frac{1}{2}} = \Delta f_0$ and from (3.64) $\delta^2 f_0 = \Delta^2 f_{-1}$. In general, it may be shown that

$$\delta^{2j+1} f_{\frac{1}{2}} = \Delta^{2j+1} f_{-j},$$
$$\delta^{2j} f_0 = \Delta^{2j} f_{-j}, \tag{3.90}$$

and it follows that Gauss's forward difference formula has the form

$$f_z \approx f_0 + \binom{z}{1}\delta f_{\frac{1}{2}} + \binom{z}{2}\delta^2 f_0 + \binom{z+1}{3}\delta^3 f_{\frac{1}{2}} + \cdots$$
$$+ \binom{z+j-1}{2j}\delta^{2j} f_0 + \binom{z+j}{2j+1}\delta^{2j+1} f_{\frac{1}{2}} + \cdots. \tag{3.91}$$

In a completely analogous manner, on introducing an interpolatory polynomial through tabular values at $z = 0, -1, 1, -2, 2, \ldots$, Gauss's backward difference formula is obtained

$$f_z \approx f_0 + \binom{z}{1}\delta f_{-\frac{1}{2}} + \binom{z+1}{2}\delta^2 f_0 + \binom{z+1}{3}\delta^3 f_{-\frac{1}{2}} + \cdots$$
$$+ \binom{z+j}{2j}\delta^{2j} f_0 + \binom{z+j}{2j+1}\delta^{2j+1} f_{-\frac{1}{2}} + \cdots. \tag{3.92}$$

Note that (3.91) and (3.92) use the central differences clustered around the horizontal line indicated in Table 3.5 at the center of the interval.

Stirling's formula is formed from the, arithmetic mean of the two Gauss formulas and for the odd-order central differences this requires a definition of the mean central difference as

$$\mu \delta f_0 = \frac{1}{2}\left(\delta f_{\frac{1}{2}} + \delta f_{-\frac{1}{2}}\right) \tag{3.93}$$

or, in general,

$$\mu\delta^n f_i = \frac{1}{2}\left(\delta^n f_{i+\frac{1}{2}} + \delta^n f_{i-\frac{1}{2}}\right),$$
$$= \frac{1}{2}\left(\delta^{n-1} f_{i+1} - \delta^{n-1} f_{i-1}\right). \quad (3.94)$$

This is a matter of convenience because Table 3.5 shows that otherwise only even-order central differences have integer suffixes.

Stirling's formula is then given as

$$f_z \approx f_0 + \binom{z}{1}\mu\delta f_0 + \frac{z}{2}\binom{z}{1}\delta^2 f_0 + \binom{z+1}{3}\mu\delta^3 f_0 + \cdots$$
$$+ \frac{z}{2j}\binom{z+j-1}{2j-1}\delta^{2j} f_0 + \binom{z+j}{2j+1}\mu\delta^{2j+1} f_0 + \cdots, \quad (3.95)$$

and on applying the definitions of (3.71), (3.72) for the factorial polynomial, it follows that

$$f_z \approx f_0 + z\mu\delta f_0 + \frac{z^2}{2!}\delta^2 f_0 + \frac{z(z^2-1^2)}{3!}\mu\delta^3 f_0 + \frac{z^2(z^2-1^2)}{4!}\delta^4 f_0 + \cdots$$
$$+ \frac{z^2(z^2-1^2)\cdots[z^2-(j-1)^2]}{(2j)!}\delta^{2j} f_0$$
$$+ \frac{z(z^2-1^2)\cdots[z^2-j^2]}{(2j+1)!}\mu\delta^{2j+1} f_0 + \cdots. \quad (3.96)$$

Stirling's formula has a smaller error term when compared with some other difference approximants to interpolated values. From (3.41) and the discussion in Sect. 3.5, it follows that an estimate of the truncation error in a polynomial of degree n is given by the leading term in the degree $n+1$ interpolate. Alternatively, if no more tabular values can be added, the error term may be estimated from knowledge of the Taylor expansion truncation error (see Sect. 2.3) and the difference values. Precise forms for the truncation error in Stirling's formula is given in the final section of this chapter after a detailed practical application.

3.11 Application of Stirling's Formula

In this section, and the Fortran application, explicit expressions are derived that go beyond what is to be found in [2] to enable a high degree of numerical precision. The practical utility of Stirling's formula is in computing interpolates or derivatives at any value of z in (3.70) in the interval $[z_{-n}, z_{+n}]$, contained by the end points, or at a specific tabular value such as $z = 0$. To consider the general case first and for later reference, we give the explicit form of (3.96) to the 12th central difference

3.11 Application of Stirling's Formula

$$f_z = f_0 + z\mu\delta f_0 + \frac{1}{2!}z^2\delta^2 f_0 + \frac{1}{3!}\left(z^3 - z\right)\mu\delta^3 f_0 + \frac{1}{4!}\left(z^4 - z^2\right)\delta^4 f_0$$

$$+ \frac{1}{5!}\left(z^5 - 5z^3 + 4z\right)\mu\delta^5 f_0 + \frac{1}{6!}\left(z^6 - 5z^4 + 4z^2\right)\delta^6 f_0$$

$$+ \frac{1}{7!}\left(z^7 - 14z^5 + 49z^3 - 36z\right)\mu\delta^7 f_0$$

$$+ \frac{1}{8!}\left(z^8 - 14z^6 + 49z^4 - 36z^2\right)\delta^8 f_0$$

$$+ \frac{1}{9!}\left(z^9 - 30z^7 + 273z^5 - 820z^3 + 576z\right)\mu\delta^9 f_0$$

$$+ \frac{1}{10!}\left(z^{10} - 30z^8 + 273z^6 - 820z^4 + 576z^2\right)\delta^{10} f_0$$

$$+ \frac{1}{11!}\left(z^{11} - 55z^9 + 1023z^7 - 7645z^5 + 21076z^3 - 14400z\right)\mu\delta^{11} f_0$$

$$+ \frac{1}{12!}\left(\begin{array}{c}z^{12} - 55z^{10} + 1023z^8 - 7645z^6 \\ + 21076z^4 - 14400z^2\end{array}\right)\delta^{12} f_0, \tag{3.97}$$

and all differences in this expression can be expressed in terms of the tabular values by repeated application of (3.61)–(3.63). The odd powers of the central difference give function values at the half-integer index points and this is the reason why the mean difference is used, as in (3.93) and (3.94). The details are shown for the first few difference terms arranged in expressions suitable for computation. For higher order differences only, the final result is given.

$$\mu\delta f_0 = \tfrac{1}{2}\left(\delta f_{\frac{1}{2}} + \delta f_{-\frac{1}{2}}\right),$$
$$= \tfrac{1}{2}(f_1 - f_{-1}), \tag{3.98}$$

$$\delta^2 f_0 = (f_1 + f_{-1}) - 2f_0, \tag{3.99}$$

$$\mu\delta^3 f_0 = \tfrac{1}{2}\left(\delta^3 f_{\frac{1}{2}} + \delta^3 f_{-\frac{1}{2}}\right),$$
$$= \tfrac{1}{2}\left[\delta^2(f_1 - f_0) + \delta^2(f_0 - f_{-1})\right],$$
$$= \tfrac{1}{2}\left[\delta\left(f_{\frac{3}{2}} - 2f_{\frac{1}{2}} + f_{-\frac{1}{2}}\right) + \delta\left(f_{\frac{1}{2}} - 2f_{-\frac{1}{2}} + f_{-\frac{3}{2}}\right)\right],$$
$$= \tfrac{1}{2}(f_2 - f_{-2}) - (f_1 - f_{-1}), \tag{3.100}$$

$$\delta^4 f_0 = (f_2 + f_{-2}) - 4(f_1 + f_{-1}) + 6f_0, \tag{3.101}$$

$$\mu\delta^5 f_0 = \tfrac{1}{2}(f_3 - f_{-3}) - 2(f_2 - f_{-2}) + \tfrac{5}{2}(f_1 - f_{-1}), \tag{3.102}$$

$$\delta^6 f_0 = (f_3 + f_{-3}) - 6(f_2 + f_{-2}) + 15(f_1 + f_{-1}) - 20 f_0, \qquad (3.103)$$

$$\mu\delta^7 f_0 = \tfrac{1}{2}(f_4 - f_{-4}) - 3(f_3 - f_{-3}) + 7(f_2 - f_{-2}) - 7(f_1 - f_{-1}), \qquad (3.104)$$

$$\delta^8 f_0 = (f_4 + f_{-4}) - 8(f_3 + f_{-3}) + 28(f_2 + f_{-2})$$
$$- 56(f_1 + f_{-1}) + 70 f_0, \qquad (3.105)$$

$$\mu\delta^9 f_0 = \tfrac{1}{2}(f_5 - f_{-5}) - 4(f_4 - f_{-4}) + \tfrac{27}{2}(f_3 - f_{-3})$$
$$- 24(f_2 - f_{-2}) + 21(f_1 - f_{-1}), \qquad (3.106)$$

$$\delta^{10} f_0 = (f_5 + f_{-5}) - 10(f_4 + f_{-4}) + 45(f_3 + f_{-3})$$
$$- 120(f_2 + f_{-2}) + 210(f_1 + f_{-1}) - 252 f_0, \qquad (3.107)$$

$$\mu\delta^{11} f_0 = \tfrac{1}{2}(f_6 - f_{-6}) - 5(f_5 - f_{-5}) + 22(f_4 - f_{-4})$$
$$- 55(f_3 - f_{-3}) + \tfrac{165}{2}(f_2 - f_{-2}) - 66(f_1 - f_{-1}), \qquad (3.108)$$

$$\delta^{12} f_0 = (f_6 + f_{-6}) - 12(f_5 + f_{-5}) + 66(f_4 + f_{-4}) - 220(f_3 + f_{-3})$$
$$+ 495(f_2 + f_{-2}) - 792(f_1 + f_{-1}) + 924 f_0. \qquad (3.109)$$

These last two results may be simplified even further by collecting the constant coefficients of each tabular value f_i. In the case of even-order central differences, the expansion in tabular function values is exactly the analogue of (3.66) and

$$\delta^{2j} f_0 = \sum_{m=0}^{2j} (-1)^m \binom{2j}{m} f_{j-m}, \qquad (3.110)$$

while the generalization to odd or even-order central differences at any tabular value f_i is

$$\delta^m f_i = \sum_{k=0}^{m} (-1)^k \binom{m}{k} f_{\frac{m+2i}{2}-k}. \qquad (3.111)$$

3.11 Application of Stirling's Formula

These results follow from application of (3.94) and the definitions in Sect. 3.8, leading to an expansion for the mean central difference

$$\mu\delta^{2j+1} f_0 = \frac{1}{2}\left(\delta^{2j+1} f_{\frac{1}{2}} + \delta^{2j+1} f_{-\frac{1}{2}}\right)$$

$$= \frac{1}{2}\left[\sum_{k=0}^{2j}\left\{(-1)^k \binom{2j+1}{k} + (-1)^{k+1}\binom{2j+1}{k+1}\right\} f_{j-k}\right]. \quad (3.112)$$

From the explicit expression (3.97), the derivative, with respect to x, of any order may be interpolated. Explicit expressions follow on observing that from $x = x_0 + zh$, it follows that

$$\frac{df(x)}{dx} = \frac{dz}{dx}\frac{df_z}{dz},$$

$$= \frac{1}{h}\frac{df_z}{dz}. \quad (3.113)$$

As an example, for the first derivative, differentiation of (3.97) gives

$$\frac{df_z}{dz} = \mu\delta f_0 + z\delta^2 f_0 + \frac{1}{3!}\left(3z^2 - 1\right)\mu\delta^3 f_0 + \frac{1}{4!}\left(4z^3 - 2z\right)\delta^4 f_0$$

$$+ \frac{1}{5!}\left(5z^4 - 15z^2 + 4\right)\mu\delta^5 f_0 + \frac{1}{6!}\left(6z^5 - 20z^3 + 8z\right)\delta^6 f_0$$

$$+ \frac{1}{7!}\left(7z^6 - 70z^4 + 147z^2 - 36\right)\mu\delta^7 f_0$$

$$+ \frac{1}{8!}\left(8z^7 - 84z^5 + 196z^3 - 72z\right)\delta^8 f_0$$

$$+ \frac{1}{9!}\left(9z^8 - 210z^6 + 1365z^4 - 2460z^2 + 576\right)\mu\delta^9 f_0$$

$$+ \frac{1}{10!}\left(10z^9 - 240z^7 + 1638z^5 - 3280z^3 + 1152z\right)\delta^{10} f_0$$

$$+ \frac{1}{11!}\left(11z^{10} - 495z^8 + 7161z^6 - 38225z^4 + 63228z^2 - 14400\right)\mu\delta^{11} f_0$$

$$+ \frac{1}{12!}\left(\begin{array}{c}12z^{11} - 550z^9 + 8184z^7 - 45870z^5 \\ +84304z^3 - 28800z\end{array}\right)\delta^{12} f_0, \quad (3.114)$$

and differentiation a second time gives the second derivative

$$\frac{d^2 f_z}{dz^2} = \delta^2 f_0 + z\mu\delta^3 f_0 + \frac{1}{4!}\left(12z^2 - 2\right)\delta^4 f_0$$

$$+ \frac{1}{5!}\left(20z^3 - 30z\right)\mu\delta^5 f_0 + \frac{1}{6!}\left(30z^4 - 60z^2 + 8\right)\delta^6 f_0$$

$$+ \frac{1}{7!}\left(42z^5 - 280z^3 + 294z\right)\mu\delta^7 f_0$$

$$+ \frac{1}{8!}\left(56z^6 - 420z^4 + 588z^2 - 72\right)\delta^8 f_0$$

$$+ \frac{1}{9!}\left(72z^7 - 1260z^5 + 5460z^3 - 4920z\right)\mu\delta^9 f_0$$

$$+ \frac{1}{10!}\left(90z^8 - 1680z^6 + 8190z^4 - 9840z^2 + 1152\right)\delta^{10} f_0$$

$$+ \frac{1}{11!}\left(110z^9 - 3960z^7 + 42966z^5 - 152900z^3 + 126456z\right)\mu\delta^{11} f_0$$

$$+ \frac{1}{12!}\begin{pmatrix} 132z^{10} - 4950z^8 + 57288z^6 - 229350z^4 \\ +252912z^2 - 28800 \end{pmatrix}\delta^{12} f_0, \quad (3.115)$$

where

$$\frac{d^2 f(x)}{dx^2} = \frac{1}{h^2}\frac{d^2 f_z}{dz^2}. \quad (3.116)$$

This process may be continued for higher order derivatives. When $z \neq i$, then there is no alternative to applying (3.114) or (3.115) directly and evaluation of the coefficient polynomials in z is required when the values of z are not known *a priori*. If the values of z are regularly spaced and recur frequently, then the corresponding coefficients should be evaluated and stored in tables.

A gross simplification of (3.114) or (3.115) occurs if $z = i$, or specifically, in the symmetric case, $z = 0$, when the interpolate for the first derivative is

$$\left(\frac{df(x)}{dx}\right)_{x=x_0} = \frac{1}{h}\begin{pmatrix} \mu\delta f_0 - \frac{1}{3!}\mu\delta^3 f_0 + \frac{4}{5!}\mu\delta^5 f_0 - \frac{36}{7!}\mu\delta^7 f_0 \\ + \frac{576}{9!}\mu\delta^9 f_0 - \frac{14400}{11!}\mu\delta^{11} f_0 \end{pmatrix}, \quad (3.117)$$

and that for the second derivative is

$$\left(\frac{d^2 f(x)}{dx^2}\right)_{x=x_0} = \frac{1}{h^2}\begin{pmatrix} \delta^2 f_0 - \frac{2}{4!}\delta^4 f_0 + \frac{8}{6!}\delta^6 f_0 - \frac{72}{8!}\delta^8 f_0 \\ + \frac{1152}{10!}\delta^{10} f_0 - \frac{28800}{12!}\delta^{12} f_0 \end{pmatrix}. \quad (3.118)$$

These results were obtained by explicit evaluation of the polynomial coefficient terms in (3.97). An alternative approach in deriving the interpolates for the derivative of any order at $z = 0$ is now described. Inspection of (3.96) shows that the polynomial coefficient terms are constituted of products of individual terms quadratic in z. Therefore, the chain rule for differentiation may be applied and, for the first derivative for the product of functions $u(z)v(z)w(z)\cdots y(z)$, the result is

$$\frac{d}{dz}\{u(z)v(z)w(z)\cdots y(z)\} = \frac{du(z)}{dz}v(z)w(z)\cdots y(z)$$

$$+ u(z)\frac{dv(z)}{dz}w(z)\cdots y(z)$$

$$\vdots$$

$$+ u(z)v(z)w(z)\cdots \frac{dy(z)}{dz}, \quad (3.119)$$

3.11 Application of Stirling's Formula

and for the second derivative

$$\begin{aligned}\frac{d^2}{dz^2}\{u(z)v(z)w(z)\cdots y(z)\} =& \frac{d^2u(z)}{dz^2}v(z)w(z)\cdots y(z)\\ &+\frac{du(z)}{dz}\frac{dv(z)}{dz}w(z)\cdots y(z)\\ &+\frac{du(z)}{dz}v(z)\frac{dw(z)}{dz}\cdots y(z)\\ &\vdots\\ &+\frac{du(z)}{dz}\frac{dv(z)}{dz}w(z)\cdots y(z)\\ &+u(z)\frac{d^2v(z)}{dz^2}w(z)\cdots y(z)\\ &+u(z)\frac{dv(z)}{dz}\frac{dw(z)}{dz}\cdots y(z)\\ &\vdots\\ &+u(z)v(z)w(z)\cdots\frac{d^2y(z)}{dz^2}.\end{aligned} \quad (3.120)$$

Now consider application of the chain rule for differentiation to the even and odd terms of (3.96)

$$\left[\frac{d}{dz}\left\{\frac{z^2(z^2-1^2)\cdots[z^2-(j-1)^2]}{(2j)!}\delta^{2j}f_0\right\}\right]_{z=0} = 0, \quad (3.121)$$

and

$$\left[\frac{d}{dz}\left\{\frac{z(z^2-1^2)\cdots[z^2-j^2]}{(2j+1)!}\mu\delta^{2j+1}f_0\right\}\right]_{z=0} = \frac{(-1^2)\cdots(-j^2)}{(2j+1)!}\mu\delta^{2j+1}f_0, \quad (3.122)$$

whereas for the second derivative

$$\left[\frac{d^2}{dz^2}\left\{\frac{z^2(z^2-1^2)\cdots[z^2-(j-1)^2]}{(2j)!}\delta^{2j}f_0\right\}\right]_{z=0} = \frac{2(-1^2)\cdots[-(j-1)^2]}{(2j)!}\delta^{2j}f_0, \quad (3.123)$$

and

$$\left[\frac{d^2}{dz^2}\left\{\frac{z(z^2-1^2)\cdots[z^2-j^2]}{(2j+1)!}\mu\delta^{2j+1}f_0\right\}\right]_{z=0} = 0, \quad (3.124)$$

so that expressions for the interpolatory approximation for first and second derivatives at $z = 0$ to any order are

$$\left(\frac{df(x)}{dx}\right)_{x=x_0} = \frac{1}{h}\left(\begin{array}{c}\mu\delta f_0 - \frac{1}{3!}\mu\delta^3 f_0 + \cdots \\ +(-1)^j \frac{1^2\cdots j^2}{(2j+1)!}\mu\delta^{2j+1} f_0 + \cdots\end{array}\right), \quad (3.125)$$

and

$$\left(\frac{d^2 f(x)}{dx^2}\right)_{x=x_0} = \frac{1}{h^2}\left(\begin{array}{c}\delta^2 f_0 - \frac{2}{4!}\delta^4 f_0 + \cdots \\ +(-1)^{j-1}\frac{2\cdot 1^2\cdots(j-1)^2}{(2j)!}\delta^{2j} f_0 + \cdots\end{array}\right). \quad (3.126)$$

A series of $(2n + 1)$-point differentiation formulas are obtained on replacing differences in (3.122), (3.123) with the respective expansions (3.112), (3.110) and collecting terms that make up the coefficients of the function values f_i

$$\left(\frac{df(x)}{dx}\right)^{<n>}_{x=x_0} \approx \frac{1}{h}\sum_{j=1}^{n} b^n_j \left(f_{-j} - f_j\right), \quad (3.127)$$

and

$$\left(\frac{d^2 f(x)}{dx^2}\right)^{<n>}_{x=x_0} \approx \frac{1}{h^2}\left[c^n_0 f_0 + \sum_{j=1}^{n} c^n_j \left(f_{-j} + f_j\right)\right]. \quad (3.128)$$

Examples of how values for the coefficients b^n_j are computed from (3.125) for $n = 1, 2$ and 3 are as follows:

$$\left(\frac{df(x)}{dx}\right)^{<1>}_{x=x_0} = \frac{1}{h}\mu\delta f_0,$$

$$= \frac{1}{2h}(f_1 - f_{-1}), \quad (3.129)$$

$$\left(\frac{df(x)}{dx}\right)^{<2>}_{x=x_0} = \frac{1}{h}\left[\mu\delta f_0 - \frac{1}{3!}\mu\delta^3 f_0\right],$$

$$= \frac{1}{h}\left\{\frac{1}{2}(f_1 - f_{-1}) - \frac{1}{3!}\left[\frac{1}{2}(f_2 - f_{-2}) - (f_1 - f_{-1})\right]\right\},$$

$$= \frac{1}{h}\left\{\frac{2}{3}(f_1 - f_{-1}) - \frac{1}{12}(f_2 - f_{-2})\right\}, \quad (3.130)$$

$$\left(\frac{df(x)}{dx}\right)^{<3>}_{x=x_0} = \frac{1}{h}\left\{\mu\delta f_0 - \frac{1}{3!}\mu\delta^3 f_0 + \frac{2^2}{5!}\mu\delta^5 f_0\right\},$$

$$= \frac{1}{h}\left\{\begin{array}{l}\frac{1}{2}(f_1 - f_{-1}) - \frac{1}{3!}\left[\frac{1}{2}(f_2 - f_{-2}) - (f_1 - f_{-1})\right] \\ +\frac{2^2}{5!}\left[\frac{1}{2}(f_3 - f_{-3}) - 2(f_2 - f_{-2}) + \frac{5}{2}(f_1 - f_{-1})\right]\end{array}\right\},$$

$$= \frac{1}{h} \left\{ \begin{array}{c} \frac{3}{4}(f_1 - f_{-1}) \\ -\frac{3}{20}(f_2 - f_{-2}) + \frac{1}{60}(f_3 - f_{-3}) \end{array} \right\}. \tag{3.131}$$

For the second derivative formulae, the c_j^n coefficients in (3.128) are computed for $n = 1$ and 2, as

$$\left(\frac{d^2 f(x)}{dx^2}\right)_{x=x_0}^{<1>} = \frac{1}{h^2} \left\{\delta^2 f_0\right\},$$

$$= \frac{1}{h^2} \left\{-2f_0 + (f_1 + f_{-1})\right\}, \tag{3.132}$$

$$\left(\frac{d^2 f(x)}{dx^2}\right)_{x=x_0}^{<2>} = \frac{1}{h^2} \left\{\delta^2 f_0 - \frac{2}{4!}\delta^4 f_0\right\},$$

$$= \frac{1}{h^2} \left\{-2f_0 + (f_1 + f_{-1}) - \frac{1}{12}\left[(f_2 + f_{-2}) - 4(f_1 + f_{-1}) + 6f_0\right]\right\},$$

$$= \frac{1}{h^2} \left\{-\frac{5}{2}f_0 + \frac{4}{3}(f_1 + f_{-1}) - \frac{1}{12}(f_2 + f_{-2})\right\}. \tag{3.133}$$

Table 3.6 shows values of the coefficients b_j^n, c_j^n for 16 decimal precision cases with the exponent of 10 shown in parentheses. The accompanying Fortran code STIR example computes extended precision results with a slightly different notation. Note that in applications, to avoid loss of precision, the choice of n should be such that $2n + 1$ does not exceed the number of significant digits available from machine precision.

3.12 Error in Stirling's Formula

This section shows expressions for the error in Stirling's $(2n+1)$-point formula for the first and second derivative in the symmetric case of (3.60). Define $\xi \in [x_{-n}, x_n]$, where ξ depends on x, so that $\xi \equiv \xi(x)$, then, from (3.36), for a $(2n+1)$-point Stirling formula interpolation (3.97), the error function on $x \in [x_{-n}, x_n]$ is defined as

$$E(x) = f(x) - p_{2n}(x)$$

$$= \frac{f^{(2n+1)}(\xi)}{(2n+1)!} \prod_{j=-n}^{+n} (x - x_j). \tag{3.134}$$

Table 3.6 Derivatives

n	j	First order $-b_j^n$	Second order c_j^n
1	0	–	−2.0000000000000000
	1	+0.5000000000000000	+1.0000000000000000
2	0	–	−2.5000000000000000
	1	+0.6666666666666666	+1.3333333333333333
	2	−0.8333333333333333(−1)	−0.8333333333333333(−1)
3	0	–	−0.2722222222222222(−1)
	1	+0.7500000000000000	+1.5000000000000000
	2	−0.1500000000000000	−0.1500000000000000
	3	+0.1666666666666667(−1)	+0.1111111111111111(−1)
4	0	–	−2.8472222222222222
	1	+0.8000000000000000	+1.6000000000000000
	2	−0.2000000000000000	−0.2000000000000000
	3	+0.3809523809523810(−1)	+0.2539682539682540(−1)
	4	−0.3571428571428571(−2)	−0.1785714285714286(−2)
5	0	–	−2.9272222222222222
	1	+0.8333333333333333	+1.6666666666666666
	2	−0.2380952380952381	−0.2380952380952381
	3	+0.5952380952380952(−1)	+0.3968253968253968(−1)
	4	−0.9920634920634921(−2)	−0.4960317460317460(−2)
	5	+0.7936507936507937(−3)	+0.3174603174603175(−3)
6	0	–	−2.9827777777777777
	1	+0.8571428571428571	+1.7142857142857143
	2	−0.2678571428571429	−0.2678571428571429
	3	+0.7936507936507937(−1)	+0.5291005291005291(−1)
	4	−0.1785714285714286(−1)	−0.8928571428571429(−2)
	5	+0.2597402597402597(−2)	+0.1038961038961039(−2)
	6	−0.1803751803751804(−3)	−0.6012506012506013(−4)

The approximation to the first derivative follows on differentiation of (3.134) with respect to x and rearrangement of terms to give

$$f'(x) = p'_{2n}(x) + \frac{1}{(2n+1)!} \frac{df^{(2n+1)}(\xi)}{dx} \prod_{j=-n}^{+n} (x - x_j)$$

$$+ \frac{f^{(2n+1)}(\xi)}{(2n+1)!} \frac{d}{dx} \prod_{j=-n}^{+n} (x - x_j), \qquad (3.135)$$

and, if $x = x_k$, then the second term is zero, so that

$$f'(x_k) = p'_{2n}(x_k) + \frac{f^{(2n+1)}(\xi_k)}{(2n+1)!} \prod_{j=-n, j \neq k}^{+n} (x_k - x_j), \qquad (3.136)$$

where $\xi_k \equiv \xi(x_k)$. In the case of equidistant symmetrically labeled points $-n \leq k \leq +n$, as in (3.60), the product term is simply products of multiples of the spacing h and (3.136) becomes

$$f'(x_k) = p'_{2n}(x_k) + (-1)^n h^{2n} \frac{n^2(n-1)^2 \cdots 1^2}{(2n+1)!} f^{(2n+1)}(\xi_k), \qquad (3.137)$$

for $x_0 - nh \leq \xi_k \leq x_0 + nh$.

3.12 Error in Stirling's Formula

Table 3.7 Error term

n	First order	Second order
1	$-\frac{1}{6}h^2 f^{(3)}(\xi)$	$\propto h$
2	$+\frac{1}{30}h^4 f^{(5)}(\xi)$	$\propto h^3$
3	$-\frac{1}{140}h^6 f^{(7)}(\xi)$	$\propto h^5$
4	$+\frac{1}{630}h^8 f^{(9)}(\xi)$	$\propto h^7$
5	$-\frac{1}{2772}h^{10} f^{(11)}(\xi)$	$\propto h^9$
6	$+\frac{1}{12012}h^{12} f^{(13)}(\xi)$	$\propto h^{11}$

Specific cases of the error term in the respective truncation of (3.129)–(3.131) are given in Table 3.7 where it is of order h^{2n} and h^{2n-1} for first and second derivative cases with a $(2n+1)$-point formula.

For the second derivative differentiation of (3.135) with respect to x gives

$$f''(x) = p''_{2n}(x) + \frac{1}{(2n+1)!} \frac{d^2 f^{(2n+1)}(\xi)}{dx^2} \prod_{j=-n}^{+n}(x-x_j)$$

$$+ \frac{2}{(2n+1)!} \frac{df^{(2n+1)}(\xi)}{dx} \frac{d}{dx} \prod_{j=-n}^{+n}(x-x_j)$$

$$+ \frac{f^{(2n+1)}(\xi)}{(2n+1)!} \frac{d^2}{dx^2} \prod_{j=-n}^{+n}(x-x_j), \qquad (3.138)$$

or, for $x = x_k$, as in (3.135), the second term is zero, while the third term survives and is $\propto h^{2n}$, as in (3.137), while the last term is $\propto h^{2n-1}$

$$f''(x_k) = p''_{2n}(x_k) + \frac{2}{(2n+1)!} \frac{df^{(2n+1)}(\xi)}{dx} \prod_{j=-n, j\neq k}^{+n}(x_k-x_j)$$

$$+ \frac{f^{(2n+1)}(\xi)}{(2n+1)!} \sum_{m=-n}^{+n} \sum_{l=-n, l\neq m}^{+n} \prod_{j=-n, j\neq l}^{+n}(x_k-x_j). \qquad (3.139)$$

For $n = 1$ to $n = 12$ coefficients in Table 3.6 (Sect. 3.11) are computed in the accompanying Fortran code using the slightly different notation in the code as described in the unpublished report [3] included with the code.

FORTRAN CODE STIR

3.13 Exercises

1. Construct the Newton interpolating polynomials from lowest to highest order approximating e^x from the following table of data pairs (x, y) and compute the corresponding values of each approximation at $x = 0.826$. Comment on what you observe.

x	y
0.80	2.2255
0.81	2.2479
0.82	2.2705
0.83	2.2933
0.84	2.3164
0.85	2.3396

2. Complete the divided difference table for the data in this table

x	y
0.0	0.0
0.5236	0.5
1.0472	0.866
1.5708	1.0

3. From the divided difference Table 3.2, write down the highest order Newton interpolating polynomial for $f(x)$ and $\frac{df(x)}{dx}$. Using these approximations, compute values for the function and the derivative at $x = 0.75$.
4. Construct the Lagrange polynomial interpolating the (x, y) data pairs $(-2, -15)$, $(-1, -8)$, $(0, -3)$ and find the smallest positive value of x at which the polynomial is zero. Does the polynomial have a maximum value on this interval?
5. Use the table of values in Exercise 2 to (i) construct the highest order Lagrange interpolating polynomial and (ii) use this polynomial to also approximate the function value at $x = 0.75$.
6. Compute an approximation to $ln(2)$ from the Lagrange polynomial interpolating the (x, y) data pairs and compare the result to the exact value,

x	y
1	0
4	1.3862944
6	1.7917595

3.14 Programming Problems

1. Consider the interpolation of the Runge function

$$\frac{1}{1+x^2}, \quad x \in [-5, +5]$$

   and the modified Runge function

$$\frac{e^{-x^2}}{1+x^2}, \quad x \in [-5, +5].$$

   Develop a code that interpolates these two functions by constructing the Newton interpolating polynomial. Verify the reliability of the code by reproducing the results of EXAMPLES 3.7 and 3.8.

2. Use the code developed in Problem 1, with equally spaced values of the argument x, to compute the Newton interpolating polynomial $p_n(x)$ for orders $n = 2, \ldots, 20$, or for the largest order possible with the available computer platform's word length and numerical precision. For each interpolating polynomial, compute the error curve and locate the maximum error for each order. Tabulate the maximum error and the argument for which it occurs. Plot the logarithm of the maximum error as a function of the order.

3. Repeat the procedure of Problem 2 for argument values that are multiples of the roots of the Chebyshev polynomials of the appropriate order using (2.38). Compare the plots of maximum error versus order of the polynomial for the two Runge functions with both equally spaced and Chebyshev nodes. Discuss and compare the convergence behavior you observe.

4. Perform the tasks of Problem 1 in the case that the interpolating polynomial is of the Lagrange form and compare the results with those of Problem 1.

References

1. Boole G (1860) Calculus of finite differences, 4th edn. Chelsea Publishing Company, New York, NY
2. Herriot JG (1963) Methods of mathematical analysis and computation. John Wiley and Sons, New York, NY
3. Delic G (1973) Formulae for numerical differentiation and integration. IKDA 73/8, Institut für Kernphysik, Technische Hochschule Darmstadt

Function Approximation

4.1 Measuring the Quality of Approximation

This chapter establishes some basic analytical properties for use in methods determining the quality of an approximation and describes some practical applications. While the emphasis is on a function of a single variable these ideas have simple generalizations to functions of several variables, or discrete argument values. When the quality of an approximation is to be studied some basic tools are required. These tools are used to measure the distance of the approximating function from the function being approximated. This necessitates a discussion and definition, of the concept of distance between functions that, in turn, requires some measure of the size of a function. Once these concepts have been clarified, a quantitative evaluation of the error in an approximation is possible. The next few sections introduce some basic ideas of applied functional analysis and then uses them in a statement of the general approximation problem. The chapter concludes with a description of some practically useful techniques for measuring error in series approximations and their manipulations. For introductory texts see [1–4] and for advanced references see [5–8]. The books by Natanson have interesting discussions on convergence in the mean in metric spaces.

4.2 Metric Spaces and Distance

Let x, y be the elements from a collection of objects constituting the set X. Because the discussion is general, this could be a set of points in a geometrical space, or a set of functions in a function space. In either case, it is assumed that X has an

Supplementary Information The online version contains supplementary material available at https://doi.org/10.1007/978-3-031-90178-2_4

algebraic structure and therefore elements of X satisfy the usual rules of vector algebra. Furthermore, the discussion is restricted to the case that X is a linear space for which $x + y \in X$ and $\lambda x \in X$ whenever $x, y \in X$ and $\lambda \in \mathbb{C}$, where $\mathbb{C}$ is the set of real or complex numbers. In the space X, the distance between elements x, y is measured by a functional $\rho(x, y)$ called the metric. The space X is said to be a metric space if, for any three elements, $x, y, z \in X$, the functional $\rho(x, y)$ is real valued and satisfies the metric space axioms:

AXIOM 4.2.I $\rho(x, y) \geq 0$, with $\rho(x, x) = 0$, and if $\rho(x, y) = 0$, then $x = y$,
AXIOM 4.2.II $\rho(x, y) = \rho(y, x)$,
AXIOM 4.2.III $\rho(x, y) \leq \rho(x, z) + \rho(z, y)$.

The last axiom is known as the triangular inequality. In this definition, the functional $\rho(x, y)$ measures the distance between elements x, y in the same set. However, the same set, X, may be transformed into different metric spaces by different choices for the distance function $\rho(x, y)$. As a simple example, consider the case $X = R^n$, an n-dimensional (real) Euclidean space, for which elements would be defined by the discrete finite sets $x = (\xi_1, \xi_2, \ldots, \xi_n)$, $y = (\eta_1, \eta_2, \ldots, \eta_n)$. A suitable choice of metric in $X = R^n$ is

$$\rho_2(x, y) = \left[(\xi_1 - \eta_1)^2 + (\xi_2 - \eta_2)^2 + \cdots + (\xi_n - \eta_n)^2 \right]^{\frac{1}{2}}. \tag{4.1}$$

For $n = 2$ the two-dimensional plane is R^2 and $\rho_2(x, y)$ is the usual result of Pythagoras shown in EXAMPLE 4.1. In another simple example consider the unit circle in R^2 that is defined by pairs of real numbers (ξ_1, ξ_2) such that $\xi_1^2 + \xi_2^2 = 1$. These are points on the circumference of the circle and for any two such pairs of points, $(\xi_1, \xi_2), (\eta_1, \eta_2)$, two possible choices of metric are

$$\rho_1(\xi_1 \xi_2, \eta_1 \eta_2) = \text{arc length between } (\xi_1, \xi_2) \text{ and } (\eta_1, \eta_2), \tag{4.2}$$

$$\rho_2(\xi_1 \xi_2, \eta_1 \eta_2) = \left[(\xi_1 - \eta_1)^2 + (\xi_2 - \eta_2)^2 \right]^{\frac{1}{2}}. \tag{4.3}$$

These two choices are shown in EXAMPLE 4.2 where ρ_2 is the chord of a circle and is a special case of (4.1).

The definition of the metric may then be used to make a quantitative statement on convergence in the space X. A sequence of elements $\{x_n\}_1^\infty \in X$ converges to an element x_0, called the limit of the sequence $\{x_n\}_1^\infty$, if $\rho(x_n, x_0) \to 0$ as $n \to \infty$. This result is abbreviated as

$$\lim_{n \to \infty} x_n = x_0. \tag{4.4}$$

It is a well known result of analysis that a convergent sequence has only one limit. However, observe that for the definition of (4.4) to be useful, the distance function ρ must be a continuous function of its arguments as otherwise $\rho(x_n, x_0)$ would not necessarily decrease as $n \to \infty$. The limit of this sequence, x_0, need not necessarily be contained in the set X as it exists only as the limit of a sequence. To demonstrate this point, consider the set of real numbers on the interval $(0, \infty]$ where the rounded

4.2 Metric Spaces and Distance

left parenthesis indicates that 0 is not a member of the set X. Choose the sequence

$$x_n = \frac{1}{n}, \tag{4.5}$$

then

$$\lim_{n \to \infty} x_n = 0. \tag{4.6}$$

We now define a set of functions that is important to much of the discussion in this book. If $[a, b]$ denotes the closed interval of the real line containing the end points a, b, then the space $\mathbf{C}[a, b]$ is constituted of arbitrary continuous functions of argument t such that $a \le t \le b$. Several choices of metric are possible here and one of these is the uniform metric ρ_∞ defined as the maximum of the difference of two elements $x, y \in \mathbf{C}[a, b]$, for the same argument t

$$\rho_\infty(x, y) = \max_{t \in [a,b]} |x(t) - y(t)|. \tag{4.7}$$

The triangle inequality of AXIOM 4.2.III is obtained as follows, for $x, y, z \in \mathbf{C}[a, b]$

$$\rho_\infty(x, y) = \max_{t \in [a,b]} |x(t) - z(t) + z(t) - y(t)|, \tag{4.8}$$

$$\le \max_{t \in [a,b]} [|x(t) - z(t)| + |z(t) - y(t)|], \tag{4.9}$$

$$\le \rho_\infty(x, z) + \rho_\infty(z, y). \tag{4.10}$$

In this case, AXIOM 4.2.III is self-evident, but AXIOM 4.2.I implies that whenever $\rho_\infty(x, y) = 0$ it must follow that $x(t) = y(t)$ for all $t \in [a, b]$. However, this may only be true for continuous functions and even then the result does not necessarily always follow.

Convergence as defined in (4.4) depends on the choice of metric and may hold for one choice but not for another. Conversely, if a sequence converges in one metric then it may also converge in another. Thus, in the case of (4.2), (4.3) and from EXAMPLE 4.2 it is evident that $\rho_2 \le \rho_1 \le \pi \rho_2$ and convergence in ρ_1 follows if and only if convergence is observed in ρ_2. However, in general, different metrics in $\mathbf{C}[a, b]$ have different convergence properties. The term "uniform" for the choice of metric in (4.7) relates to the nature of the convergence. A sequence $\{x_n\}_1^\infty$ that converges in the uniform metric does so uniformly. This means that, for example, in the case of a function space, whenever (4.4) is true then the choice of metric in (4.7) ensures that convergence occurs at any, or all values of the argument t. Convergence is said to occur uniformly, and furthermore the limit element $x_0(t)$ is also continuous and therefore $x_0(t) \in \mathbf{C}^\infty[a, b]$. These special convergence results do not hold for other choices of metric.

EXAMPLES 4.1–4.4

4.3 Normed Spaces and Size

To measure the size of an element x of a space X we introduce the norm $\|\cdot\|$ that is also a functional such that, when it operates on x, the result is a real number such that $\|\cdot\| \geq 0$ and $\|\cdot\| = 0$ if $\mathbf{0}$ is the null element of the set X. If $\lambda \in \mathbb{C}$, where $\mathbb{C}$ is the set of real or complex numbers, then the norm $\|\cdot\|$ must satisfy three axioms:

AXIOM 4.3.I $\| x \| = 0$, if, and only if $x = \mathbf{0}$.
AXIOM 4.3.II $\| \lambda x \| = |\lambda| \| x \|, \lambda \in \mathbb{C}$.
AXIOM 4.3.III $\| x + y \| \leq \| x \| + \| y \|$.

The last axiom is the triangular inequality of the norm and AXIOM 4.3.II expresses homogeneity of the norm. A set X endowed with a fixed norm is called a normed space. Normed spaces have special properties some of which are listed in the following.

In a normed space X the metric for $x, y \in X$ may be defined as

$$\rho(x, y) = \| x - y \|, \tag{4.11}$$

and it follows that, when $\rho(x, y) = 0$, then also $\| x - y \| = 0$, or $x - y = 0$, and $x = y$. From (4.11), the triangular inequality for the metric follows from that for the norm

$$\| x - y \| = \| (x - z) + (z - y) \|,$$
$$\leq \| x - z \| + \| z - y \|, \tag{4.12}$$

or

$$\rho(x, y) \leq \rho(x, z) + \rho(z, y). \tag{4.13}$$

Another useful inequality for the norm is the following:

$$|\| x \| - \| y \|| \leq \| x - y \|. \tag{4.14}$$

The importance of the norm is that it may be used to measure the size of elements x in X. Thus the set of elements X is said to be bounded, if and only if, a constant β exists such that, for every $x \in X$ the norm is bounded by β, or $\| x \| \leq \beta$. The norm is especially useful in studying convergence by measuring the size of the discrepancy between the limit element x_0 and some member of the sequence $\{x_n\}_1^\infty \in X$. The sequence is said to converge in norm to x_0 if

$$\| x_n - x_0 \| \to 0, \tag{4.15}$$

where "$\to$" indicates the limit $n \to \infty$. An alternative statement of (4.15) is that given an $\epsilon > 0$, then there exists an integer $N(\epsilon)$, such that

$$\| x_n - x_0 \| \leq \epsilon, \; n \geq N(\epsilon). \tag{4.16}$$

In the definition of the norm and for (4.15) to be useful, it is a requirement that the norm is a continuous function of the argument. In other words, as $\|\cdot\|$ ranges over all elements x_n in the sequence it is observed that

$$\|x_n\| \to \|x_0\| \qquad (4.17)$$

whenever $x_n \to x_0$. Note the difference in information content between (4.15) and (4.17) concerning the limit $n \to 0$. The result of (4.15) states that the difference between x_n and x_0 approaches 0, while (4.17) states that the size of x_n approaches the size of x_0. For any n the two results of (4.15), (4.17) may be related through the inequality in (4.14)

$$|\|x_n\| - \|x_0\|| \leq \|x_n - x_0\|. \qquad (4.18)$$

EXAMPLES 4.5 and 4.6

4.4 Examples of Normed Spaces

Because the emphasis in this book is on functions of either a discrete or a continuous variable, with special interest focused on the latter, examples of norms for both types of spaces are shown here.

In the finite-dimensional space $\mathbf{K}^n$ consisting of complex elements of finite discrete sets $x = (\xi_1, \xi_2, \ldots, \xi_n)$ several possible choices of norm are as follows.

The Euclidean norm

$$\|x\|_2 = \left[\sum_{i=1}^{n} |\xi_i|^2\right]^{\frac{1}{2}}, \qquad (4.19)$$

the uniform norm

$$\|x\|_\infty = \max_{i=1,\ldots,n} |\xi_i|, \qquad (4.20)$$

the ℓ^1 norm

$$\|x\|_1 = \sum_{i=1}^{n} |\xi_i|, \qquad (4.21)$$

and EXAMPLE 4.6 shows examples of these three choices of norm in $\mathbf{R}^2$. The last choice in (4.21) is a special case of the next example.

Consider the limiting case of $\mathbf{K}^n$, when the discrete index n is denumerably infinite. Each element of such a space is made up of all sequences $x = (\xi_1, \xi_2, \ldots)$, written simply as $\{\xi_i\}$ and ℓ^p is the space of all elements showing boundedness for

$$\sum_{i=1}^{\infty} |\xi_i|^p \leq \infty. \qquad (4.22)$$

For two elements $x = \{\xi_i\}$ and $y = \{\eta_i\}$ of ℓ^p, a simple application of Minkowski's[1] inequality for series shows that

$$\left[\sum_{i=1}^{\infty} |\xi_i + \eta_i|^p\right]^{\frac{1}{p}} \leq \left[\sum_{i=1}^{\infty} |\xi_i|^p\right]^{\frac{1}{p}} + \left[\sum_{i=1}^{\infty} |\eta_i|^p\right]^{\frac{1}{p}} \leq \infty, \quad (4.23)$$

and it may be concluded that $x + y$ is also in ℓ^p. A suitable norm for ℓ^p is

$$\|x\|_p = \left[\sum_{i=1}^{\infty} |\xi_i|^p\right]^{\frac{1}{p}}. \quad (4.24)$$

Then Minkowski's inequality in (4.23) is also the triangular inequality of the norm.

In the case for which the elements of the space are continuous functions of a continuous variable $t \in [a, b]$ a suitable choice of norm in $\mathbf{C}[a, b]$ is the uniform norm

$$\|x\|_\infty = \max_{t \in [a,b]} |x(t)|. \quad (4.25)$$

The analogue to ℓ^p for functions continuous on $t \in [a, b]$ is the space $\mathbf{L}^p[a, b]$ for integer $p \geq 1$, which, similar to (4.24), requires that all elements of the space are integrable on the interval so that

$$\int_a^b |x(t)|^p \, dt < \infty. \quad (4.26)$$

While functions that do not satisfy (4.26) exist, they are not elements of the space $\mathbf{L}^p[a, b]$. For integrals, Minkoswki's inequality is

$$\left[\int_a^b |x(t) + y(t)|^p \, dt\right]^{\frac{1}{p}} \leq \left[\int_a^b |x(t)|^p \, dt\right]^{\frac{1}{p}} + \left[\int_a^b |y(t)|^p \, dt\right]^{\frac{1}{p}}, \quad (4.27)$$

and a suitable norm is

$$\|x\|_p = \left[\int_a^b |x(t)|^p \, dt\right]^{\frac{1}{p}}. \quad (4.28)$$

Elements of a space $\mathbf{L}^r$ are also elements of a space $\mathbf{L}^s$, with $s < r$ and then

$$\|x\|_s \leq \|x\|_r. \quad (4.29)$$

The significance of the value of p in the norm of (4.28) is demonstrated schematically in EXAMPLE 4.8 where, for real functions $x, y \in \mathbf{C}[0, 3]$, the functional $\|x - y\|_p$ is shown for several values of p. If $p < 1$, then $\|x\|_p$ produces values that give similar large weights to a range of values for $x - y$. However, when $p > 1$ then $\|x\|_p$

---

[1] Hermann Minkowski, German mathematician, 1864–1909.

produces values that give an increasing weight as $x - y$ increases. In particular, the limit $p \to \infty$ corresponds to the uniform norm when only the maximal value of $\mid x - y \mid$ is used by the norm and all other values are ignored. The usage " uniform", as explained at the end of Sect. 4.1, applies also for the norm.

EXAMPLES 4.7 and 4.8

4.5 Hilbert Space

The previous section showed examples of how norms may be chosen for a set of elements **X** and it is obvious that the choice is not unique even for the same set of elements. However, if the norm is chosen in a special (if not unique) way, then some remarkable properties follow. In the examples of (4.24), (4.28) a special role is played by the choice of $p = 2$. This choice corresponds to the analogy from Euclidean geometry mentioned in (4.19). Hilbert[2] space is the space of general elements that is the closest analogue to this Euclidean example. Before making a definition the inner product is first introduced with the notation $(\cdot, \cdot)$ and complex conjugate $\overline{(\cdot, \cdot)}$.

A space of elements **X** is an inner product space if, for each pair $x, y \in \mathbf{X}$, a complex number called the inner product (x, y) exists and satisfies the following axioms:

AXIOM 4.5.I $(x, y) = \overline{(y, x)}$.
AXIOM 4.5.II $(\lambda x_1 + \mu x_2, y) = \lambda(x_1, y) + \mu(x_2, y)$, $\lambda, \mu \in \mathbb{C}$.
AXIOM 4.5.III $(x, x) \geq 0$, with equality only if $x = \mathbf{0}$.

Where $x = \mathbf{0}$, is the null element of the space. The complex conjugate applies to the second element of the inner product. Using the axioms it may be shown that for a linear combination in the second position of the inner product

$$(x, \lambda y_1 + \mu y_2) = \overline{\lambda}(x, y_1) + \overline{\mu}(x, y_2), \tag{4.30}$$

and for the null element

$$(x, \mathbf{0}) = (\mathbf{0}, y) = 0, \tag{4.31}$$

while the Cauchy[3] inequality for the inner product is

$$\mid (x, y) \mid^2 \leq (x, x) \cdot (y, y). \tag{4.32}$$

The space **X** becomes a normed space when a suitable norm is chosen. In particular it becomes a Hilbert space if the norm is directly defined in terms of the inner product by

$$\| x \| = \sqrt{(x, x)}, \tag{4.33}$$

for all $x \in \mathbf{X}$ and, in this case, the Hilbert space is distinguished by the notation **H**.

---

[2] David Hilbert, German mathematician, 1862–1943.
[3] Augustin Louis Cauchy, French mathematician, 1789–1857.

Two simple examples of a Hilbert space are given below for a function of a discrete and continuous variable, respectively. For the first example consider the space ℓ^2, discussed in Sect. 4.4, with elements

$$x = (\xi_1, \xi_2, \ldots), \tag{4.34}$$
$$y = (\eta_1, \eta_2, \ldots), \tag{4.35}$$

such that (4.22) holds with $p = 2$. This becomes a Hilbert space if the inner product is defined as

$$(x, y) = \sum_{i=1}^{\infty} \xi_i \overline{\eta_i}, \tag{4.36}$$

with the norm as a special case of (4.24)

$$\| x \| = \left[\sum_{i=1}^{\infty} | \xi_i |^2 \right]^{\frac{1}{2}}. \tag{4.37}$$

In (4.36) note that the complex conjugate of the discrete set in (4.35) appears because y is in the second position of the inner product. For the second example of a Hilbert space, consider the case of functions of a continuous variable. Let $\phi(t)$ be a function defined on $[a, b]$ that is integrable in the sense of (4.26) and almost everywhere positive, $\phi(t) > 0$. Then $\mathbf{L}_\phi^2[a, b]$ is the Hilbert space of functions $x(t)$ that are square integrable with respect to $\phi(t)$ on the interval $[a, b]$

$$\int_a^b | x(t) |^2 \phi(t) \, dt < \infty, \tag{4.38}$$

with the inner product defined by

$$(x, y) = \int_a^b \left[x(t) \overline{y(t)} \right] \phi(t) \, dt, \tag{4.39}$$

and the norm

$$\| x \| = \left[\int_a^b | x(t) |^2 \phi(t) \, dt \right]^{\frac{1}{2}}. \tag{4.40}$$

Again the complex conjugate of the function $y(t)$ appears in (4.39) because y is the element in the second position of the inner product. If $\phi(t) = 1$, this is a special case of the norm defined in (4.28) for $\mathbf{L}^2[a, b]$.

The presence of an inner product in Hilbert space leads to Euclidean-like concepts that are important in approximation methods and are the direct result of orthogonality which follows from the inner product as is described in the next section.

EXAMPLES 4.9 and 4.10

4.6 The Orthonormal Basis in Hilbert Space

If two elements $x, y \in \mathbf{H}$, neither of which is the null element of $\mathbf{H}$, have the result that their inner product is zero, then they are said to be orthogonal with the notation

$$x \perp y \text{ if } (x, y) = 0. \tag{4.41}$$

If $\mathbf{E} \subset \mathbf{H}$ is a subset of elements and $x \in \mathbf{H}$ is orthogonal to all elements of $\mathbf{E}$ then it is represented as $x \perp \mathbf{E}$. Furthermore, if $\mathbf{E}_1, \mathbf{E}_2$ are two subsets of $\mathbf{H}$ such that all elements from one are orthogonal with all elements of the other, then $\mathbf{E}_1 \perp \mathbf{E}_2$. In this case, the Hilbert space $\mathbf{H}$ may be decomposed into the direct sum of $\mathbf{E}_1$ and $\mathbf{E}_2$ represented symbolically as

$$\mathbf{H} = \mathbf{E}_1 \oplus \mathbf{E}_2. \tag{4.42}$$

If a set of elements $\{x_n\}_1^\infty$ in a Hilbert space is such that all members of the set are pairwise orthogonal

$$(x_i, x_j) = 0, \ i \neq j, \tag{4.43}$$

$\{x_n\}_1^\infty$ is called an orthogonal set. If, in addition, the norm of each member of the set is unity

$$\| x_i \| = 1, \ i = 1, 2, \ldots, \tag{4.44}$$

then the set is said to be orthonormal. The importance of the last type of set is that a Hilbert space, even if it is infinite-dimensional, always has a denumerable orthonormal set. Orthonormal sets are of special importance because they may be used to represent any element x of the Hilbert space $\mathbf{H}$ as a generalized Fourier series

$$x = \sum_{i=1}^{\infty} a_i x_i. \tag{4.45}$$

An expression for the coefficients a_i is obtained on multiplication of both sides in (4.45) by $\bar{x}_j$ and application of the inner product of (4.43) to give

$$(x, x_j) = \sum_{i=1}^{\infty} a_i (x_i, x_j), \tag{4.46}$$

and it then follows from orthogonality, in (4.43), that only the $i = j$ term survives on the right-hand side of (4.46). Therefore the Fourier coefficients of (4.45) are the numbers

$$a_i = (x, x_i). \tag{4.47}$$

To better understand the Euclidean character of these results consider a simple Hilbert space $\mathbf{H}_2$ that has the orthonormal set $\{x_n\}_1^2$ shown schematically in EXAMPLE 4.11. Any element $x \in \mathbf{H}_2$ has a unique representation

$$x = a_1 x_1 + a_2 x_2, \tag{4.48}$$

where

$$a_1 = (x, x_1), \qquad (4.49)$$
$$a_2 = (x, x_2). \qquad (4.50)$$

The numbers a_1, a_2 may be viewed as the " coordinates" of x along the " directions" x_1, x_2 that are " orthogonal" in the Euclidean sense. We could also consider the representation on the right-hand side of (4.48) as the result of an orthoprojector operator pair, P and Q, defined as follows:

$$P + Q = I, \qquad (4.51)$$

$$Px_1 = x_1, \qquad (4.52)$$
$$Px_2 = 0, \qquad (4.53)$$

$$Qx_1 = 0, \qquad (4.54)$$
$$Qx_2 = x_2, \qquad (4.55)$$

from which it follows that

$$Px = a_1 x_1, \qquad (4.56)$$
$$Qx = a_2 x_2, \qquad (4.57)$$

with the identity operator

$$Ix = x. \qquad (4.58)$$

Therefore it is usual to speak of a_i as the projection of x along the " direction" x_i with reference to the concept shown in EXAMPLE 4.11 and the results of (4.51)–(4.57). The interpretation of operators P, Q is that they sort the elements of $\mathbf{H}_2$ into two orthogonal subsets $\mathbf{E}_1$, $\mathbf{E}_2$. This allows $\mathbf{H}_2$ to be written in the form of (4.42) and therefore each element $x \in \mathbf{H}$ has the unique representation shown in (4.48).

In view of (4.45) and the orthogonality property, this simple idea from Euclidean geometry has an extension to any number of dimensions. Consider the subspace $\mathbf{H}_n$ that consists of all possible combinations of the first n elements of the orthonormal set $\{x_n\}_1^\infty$ in $\mathbf{H}$. Then $x \in \mathbf{H}$, has the nth partial sum of the Fourier series (4.45) as

$$S_n = \sum_{i=1}^{n} a_i x_i, \qquad (4.59)$$

which is a projection of x into the n-dimensional subspace $\mathbf{H}_n \subset \mathbf{H}$. Thus (4.59) is a general statement of the truncated expansion methods to be introduced in the following chapters. Furthermore, an explicit means of obtaining the unknown coefficients of (4.59) is given in (4.47). The choice given in (4.47) is very special and to understand why, consider an alternative set of numbers $\{\alpha_i\}_1^n$ such that $\alpha_i \neq a_i$. This

4.6 The Orthonormal Basis in Hilbert Space

alternative set of coefficients is then used to define an element $z \in \mathbf{H}_n$ as

$$z = \sum_{j=1}^{n} \alpha_j x_j. \tag{4.60}$$

Now, as a first practical application of the norm, compute the distance between x and z using the square of the norm in $\mathbf{H}_n$ to avoid carrying a square root operation throughout. Application of the axioms of the inner product and the orthogonality result in (4.43) leads to

$$\begin{aligned} \| x - z \|^2 &= (x-z, x-z) \\ &= (x,x) - (z,x) - (x,z) + (z,z) \\ &= \| x \|^2 - \sum_{j=1}^{n} \alpha_j (x_j, x) - \sum_{i=1}^{n} \overline{\alpha}_i (x, x_i) + \sum_{j=1}^{n}\sum_{i=1}^{n} \alpha_j \overline{\alpha}_i (x_j, x_i) \\ &= \| x \|^2 - \sum_{j=1}^{n} \alpha_j (x_j, x) + \sum_{i=1}^{n} \overline{\alpha}_i [\alpha_i - (x, x_i)], \end{aligned} \tag{4.61}$$

which shows that if $\alpha_i = (x, x_i)$ then the last term vanishes and the distance between x and z is at a minimum. While this distance can be made small it cannot be zero because x is defined in terms of the infinite sum of (4.45) and the choice $\alpha_i = a_i$ in (4.61) gives the result known as Bessel's identity

$$\| x - S_n \|^2 = \| x \|^2 - \sum_{i=1}^{n} | a_i |^2 . \tag{4.62}$$

From the fact that the left-hand side is non-zero there follows Bessel's inequality

$$\| x \|^2 \geq \sum_{i=1}^{n} | a_i |^2, \tag{4.63}$$

which also holds in the limit $n \to \infty$.

To conclude this section, we observe that there is another interpretation to the meaning of the partial sum S_n as defined in (4.59). Again, we introduce the idea of the orthoprojector by subdividing the Hilbert space $\mathbf{H}$ into two parts with the first containing $\mathbf{H}_n$ and the second containing all elements not in $\mathbf{H}_n$ which is represented by set notation as $\mathbf{H} - \mathbf{H}_n$. Clearly, as may be shown by direct application of orthogonality in (4.43), the two parts of Hilbert space are orthogonal and $(\mathbf{H} - \mathbf{H}_n) \perp \mathbf{H}_n$. Therefore, just as in the simple example of (4.48), it is possible to define orthoprojectors

$$P_n + Q_n = I, \tag{4.64}$$
$$P_n \mathbf{H} = \mathbf{H}_n, \tag{4.65}$$
$$Q_n \mathbf{H} = \mathbf{H} - \mathbf{H}_n. \tag{4.66}$$

A special case of (4.65) would be if in (4.59) the partial sum is obtained from

$$P_n x = S_n. \tag{4.67}$$

This discussion is the exact analogue of the decomposition of $\mathbf{H}$ in (4.42) with $\mathbf{E}_1 = \mathbf{H}_n$ and $\mathbf{E}_2 = \mathbf{H} - \mathbf{H}_n$.

EXAMPLES 4.11 and 4.12

4.7 The General Approximation Problem

As a consequence of the discussion in the previous sections, it is now possible to give a general statement of the approximation problem. In (4.59) the choice of the parameter set $A_n = \{a_i\}_1^n$ determines the distance between an exact element x and the approximation z to that element in Hilbert space $\mathbf{H}_n$, as in (4.61). In view of the discussion of (4.61) it follows that there is a choice $A_n^* = \{a_i^*\}_1^n$ that minimizes the distance $\| x - z \|$ and then the choice A_n^* represents a " best approximation" in this sense. Also, the definition of (4.59) requires a choice of basis defined on some set T. In most applications in this book this is the set of basis functions defined on the set $T = [a, b]$. For example, in the case of the discussion to follow, this could be a polynomial or trigonometric basis on T. Because the approximation in (4.45) is a linear combination of the basis elements defined on the set T and scaled by parameters from the set A_n, it may be schematically represented as a functional $F(A_n, T)$.

The approximation is a truncated form such as that in (4.59) the " best approximation" in the truncated space $\mathbf{H}_n$ is the orthogonal projection $P_n x = S_n$ of (4.67). Any other approximation chosen from $\mathbf{H}_n$ will increase the distance from the element being approximated as is seen from the discussion of (4.61). These qualitative arguments may be collected into the following statement of the general approximation problem in a Hilbert space.

Theorem 4.1 Let $\mathbf{H}$ be a Hilbert space with norm $\| \cdot \|$ and elements defined on a set T. If $\mathbf{H}_n \subset \mathbf{H}$ then $F(A_n, T) \in \mathbf{H}$ is an approximation to $x \in \mathbf{H}$ that depends on a distinct set of parameters $A_n = \{a_i\}_1^n$. The " best approximation" $F^*(A_n^*, T) \in \mathbf{H}_n$ to $x \in \mathbf{H}$ corresponds to the set A_n^* for which F^* is the element of $\mathbf{H}_n$ closest to $x \in \mathbf{H}$ and for all other choices of A

$$\| F^* - x \| < \| F - x \|. \tag{4.68}$$

Lemma 4.2 The element x is projected onto the subspace $\mathbf{H}_n$ of $\mathbf{H}$ and the error of the best approximation is the residual

$$E^* = x - F^*. \tag{4.69}$$

As a geometrical demonstration of the significance of this statement, consider again the example of $\mathbf{H}_2$ in EXAMPLE 4.13. The element being approximated is the point

4.7 The General Approximation Problem

x and all approximations lie in the subspace $\mathbf{H}_1 \subset \mathbf{H}_2$ shown as a horizontal line in EXAMPLES 4.11, 4.13. While any element in $\mathbf{H}_1$ could be used to approximate x, obviously, the best approximation, F^*, is the orthoprojection of x onto $\mathbf{H}_1$. Any other approximation $F^* + \lambda F$, $\lambda \in \mathbb{R}$, which is displaced from F^*, increases the distance between x and the approximation.

While the above discussion for EXAMPLES 4.11, 4.13 is a simple geometrical interpretation it also holds in the general case as well. Thus, as a " proof" of the above, the following shows algebraically that for F^*, $F \in \mathbf{H}_n$ the minimum with respect to the real parameter λ for the functional

$$f(\lambda) = \| x - (F^* + \lambda F) \|^2 \tag{4.70}$$

gives the result

$$(E^*, F) = 0. \tag{4.71}$$

This states that the error of the best approximation, the residual E^*, is orthogonal to any approximant $F \in \mathbf{H}_n$ and therefore orthogonal to the subspace $\mathbf{H}_n$. In other words, the residual E^* of the approximation F^* to x is always " zero" in the subspace in $\mathbf{H}_n$ in the sense that the inner product (E^*, F) is zero. This is true for each choice of the dimension of the subspace $\mathbf{H}_n$ because, on using the orthoprojectors of (4.64) to (4.66), it follows that $P_n E^* = \mathbf{0}$ and therefore $E^* \in \mathbf{H} - \mathbf{H}_n$.

To demonstrate the result of (4.71) consider the function of λ defined by (4.70), clearly $f(\lambda)$ has a minimum when $\lambda = 0$, and increases if λ ranges above or below 0. Therefore the derivative with respect to λ is minimal, consider

$$\frac{df}{d\lambda} = \frac{d}{d\lambda}(x - F^* - \lambda F, x - F^* - \lambda F), \tag{4.72}$$

$$= \frac{d}{d\lambda}\left[\| x - F^* \|^2 - 2\lambda(x - F^*, F) + \lambda^2 \| F \|^2 \right], \tag{4.73}$$

$$= -2(E^*, F) + 2\lambda \| F \|^2, \tag{4.74}$$

and equating the last result to zero corresponds to the minimum of $f(\lambda)$, thus yielding

$$\lambda = \frac{(E^*, F)}{\|F\|^2}. \tag{4.75}$$

Then the result of (4.71) follows because $\lambda = 0$ at the minimum of $f(\lambda)$. Using the result $(E^*, F) = 0$ it follows that

$$(E^*, F) = (x - F^*, F), \tag{4.76}$$
$$= (x, F) - (F^*, F), \tag{4.77}$$

or

$$(x, F) = (F^*, F). \tag{4.78}$$

Having demonstrated (4.71), then the result of (4.68) may now be derived by applying (4.78) to show

$$\| x - F \|^2 - \| x - F^* \|^2 = \| F - F^* \|^2 \qquad (4.79)$$
$$\geq 0, \qquad (4.80)$$

where the right-hand side is always seen to be greater than zero. The last result in (4.79) has a simple geometrical interpretation for $x \in \mathbf{H}$ and $F^*, F \in \mathbf{H}_n$. Because $F - F^* \in \mathbf{H}_n$ and $(x - F^*) \perp \mathbf{H}_n$, then $x - F$ is the hypotenuse of a right-angled triangle and, in the Hilbert space $\mathbf{H}$, (4.79) is the result of Pythagoras for such a triangle. This discussion is demonstrated for the $n = 2$ case in EXAMPLE 4.13 where the subspace $\mathbf{H}_1$ is a line. These results hold for any finite n and the analogy demonstrates again that the geometrical interpretation is typical of the Euclidean-like character of Hilbert space.

EXAMPLE 4.13

4.8 The L^2 Error and Truncated Series Expansions

The consequences of the preceding discussion are that a concise and practical method may be constructed to give a precise measure of the approximation error due to truncation of a series representation in the form of (2.13). Furthermore, convergence properties may be studied with this method to complement and extend the discussion of Sect. 2.10.

Consider the application of results discussed in this chapter to the space $\mathbf{H} = \mathbf{L}^2[a, b]$ of real functions $x(t)$ continuous on the interval $t \in [a, b]$ with the norm $\| x \| = \sqrt{(x, x)}$ and inner product

$$(x, y) = \int_a^b x(t) y(t) \varphi(t) dt \qquad (4.81)$$

for a real weight function $\varphi(t) > 0$ on $t \in [a, b]$. In this section we consider again the possible choices of orthogonal basis sets introduced in Sects. 2.4 and 2.5. An expression is derived for the error of truncated series using these bases by applying the definitions of the norm in the Hilbert space $\mathbf{L}^2[a, b]$.

In Chap. 2 it was shown that the Legendre series realization of the general form in (4.45) for a function $f(t), t \in [+1, -1]$ is

$$f(t) = \sum_{k=0}^{\infty} (k + \frac{1}{2}) g_k P_k(t), \qquad (4.82)$$

4.8 The L^2 Error and Truncated Series Expansions

with coefficients analogous to those of (4.47) and defined in (2.52). Similarly for the Chebyshev case the general form (4.45) is the series

$$f(t) = \sum_{k=0}^{\infty\prime} h_k T_k(t), \qquad (4.83)$$

where the prime indicates that the first term of the series has included a factor of $\frac{1}{2}$ and reads $\frac{1}{2}h_0 T_0(x)$ while the coefficients of (4.47) are defined in (2.54). When the series in (4.82), (4.83) are truncated at $k = N$, the $(N+1)$-dimensional Hilbert spaces spanned by the first $N+1$ polynomials P_k or T_k, respectively, have the inner product of (4.81) with the weight function for Legendre and Chebyshev cases, respectively, as

$$\varphi(t) = 1, \qquad (4.84)$$

$$\varphi(t) = \frac{1}{\sqrt{1-t^2}}. \qquad (4.85)$$

However, if the function to be approximated is known to be periodic, such that for a fixed period, $T > 0$, when $f(t+T) = f(t)$, then the Fourier series introduced in Sect. 2.5 are usually applied. In this case an appropriate choice of basis functions is the trigonometric functions, such as sine and cosine. As discussed in Chap. 2, an arbitrary function $f(t)$ may be expanded in Fourier series in argument $t \in [+1, -1]$

$$f(t) = \sum_{k=0}^{\infty\prime} c_k \cos(k\pi t) + \sum_{k=1}^{\infty} s_k \sin(k\pi t) \qquad (4.86)$$

with the basis functions

$$\cos(0\pi t), \sin(1\pi t), \cos(1\pi t), \ldots, \sin(k\pi t), \cos(k\pi t), \ldots, \qquad (4.87)$$

where, as before, the prime indicates that the first term has included a factor of $\frac{1}{2}$ and it reads $\frac{1}{2}c_0 \cos(0\pi t)$. On introduction of the variable $\theta = \pi t$ the Hilbert space spanned by the set of elements in (4.87) has an inner product (4.81) defined in terms of either variable t or θ with a weight function of unity

$$(y, z) = \frac{1}{\pi} \int_{-\pi}^{+\pi} y(\theta) z(\theta) \, d\theta, \qquad (4.88)$$

$$= \int_{-1}^{+1} y(t) z(t) \, dt. \qquad (4.89)$$

From the discussion of (4.62) we now obtain a simple expression, called the L^2 error for the distance between the approximant and the exact result in the Hilbert space $L^2[a, b]$. For example, in the Legendre case, if f_N^P denotes the approximant obtained when the series for $f(t)$ in (4.82) is terminated at the $k = N$ term, then the L_P^2 error is defined as

$$L_P^2(N) = ||f - f_N^P||. \tag{4.90}$$

The simple and practically useful expression for this result follows from the definition in (4.33) of the norm in Hilbert space

$$\begin{aligned}L_P^2(N) &= \sqrt{\left(f - f_N^P, f - f_N^P\right)}, \\ &= \sqrt{\{(f, f) - 2(f_N^P, f) + (f_N^P, f_N^P)\}},\end{aligned} \tag{4.91}$$

and, in this last expression, the separate terms are obtained by application of the orthogonality property of the Legendre polynomials

$$\begin{aligned}(f, f) &= \sum_{k=0}^{\infty}\sum_{j=0}^{\infty}(k+\tfrac{1}{2})(j+\tfrac{1}{2})g_k g_j (P_k, P_j), \\ &= \sum_{k=0}^{\infty}(k+\tfrac{1}{2})g_k^2,\end{aligned} \tag{4.92}$$

and also

$$\begin{aligned}(f_N^P, f_N^P) &= \sum_{k=0}^{N}(k+\tfrac{1}{2})g_k^2, \\ &= (f_N^P, f).\end{aligned} \tag{4.93}$$

Therefore, from (4.91) to (4.93), it follows that

$$\begin{aligned}L_P^2(N) &= ||f - f_N^P||, \\ &= \left[\sum_{k=N+1}^{\infty}(k+\tfrac{1}{2})g_k^2\right]^{\tfrac{1}{2}}.\end{aligned} \tag{4.94}$$

Likewise, let f_N^T, f_N^F denote the approximants obtained when the series for $f(t)$ in (4.83) and (4.86), respectively, are terminated at the $k = N$ term. A similar method gives the L^2 error for the Chebyshev and Fourier cases, respectively, as

$$\begin{aligned}L_T^2(N) &= ||f - f_N^T||, \\ &= \left[\frac{\pi}{2}\sum_{k=N+1}^{\infty}h_k^2\right]^{\tfrac{1}{2}},\end{aligned} \tag{4.95}$$

$$\begin{aligned}L_F^2(N) &= ||f - f_N^F||, \\ &= \left[\sum_{k=N+1}^{\infty}\left(c_k^2 + s_k^2\right)\right]^{\tfrac{1}{2}},\end{aligned} \tag{4.96}$$

4.8 The L^2 Error and Truncated Series Expansions

and in the Fourier case

$$(f, f) = (\frac{1}{2}c_0)^2 + \sum_{k=1}^{\infty} c_k^2 + \sum_{k=1}^{\infty} s_k^2, \qquad (4.97)$$

with

$$(f_N^F, f_N^F) = (\frac{1}{2}c_0)^2 + \sum_{k=1}^{N} c_k^2 + \sum_{k=1}^{N} s_k^2,$$
$$= (f_N^F, f). \qquad (4.98)$$

While an alternative and simpler method of obtaining the results of (4.94)–(4.96) is by direct application of Bessel's identity (4.62), the steps outlined above show how the norm is applied in manipulations. In numerical applications the L^2 error is easily computed from the expressions in (4.94) to (4.96) and it is an important numerical quantity because it measures the error in the norm of the corresponding Hilbert space. These expressions should be compared to the bound of (2.174) that does not correspond to a normed Hilbert space. Therefore, the L^2 error is the preferred measure of convergence properties of series expansions. But nevertheless, an $\mathbf{L}^2$ norm gives a different measure from that of the $\mathbf{L}^\infty$ case. The latter choice is the most precise method of determining the accuracy of the approximation for all three expansion basis sets, with N sufficiently large, because it inspects the respective error curves

$$E_N^P(t) = f(t) - f_N^P(t), \qquad (4.99)$$

$$E_N^T(t) = f(t) - f_N^T(t), \qquad (4.100)$$

$$E_N^F(t) = f(t) - f_N^F(t), \qquad (4.101)$$

to determine the behavior of the maximum error as $N \to \infty$. However, determination of the quality of the approximation by inspection of the error curve requires values of both the function being approximated and the approximant at all values of the argument t on the interval of approximation. On the other hand, the L^2 error does not depend on the argument t explicitly because it is defined in terms of an integral (4.81) over the variable t. Obviously, these integrals do not have to be evaluated as the L^2 errors may be computed by replacing infinity in the summations of (4.94)–(4.96) with a suitable N_{max} and then computing the $L^2(N)$ error for each value of the range $N = 1, \ldots, N_{max} - 1$. Graphs of log_{10} of $L_P^2(N)$, $L_T^2(N)$, $L_F^2(N)$, as functions of N measure the progress in accuracy of the approximant as more terms are added to the respective series. If such graphs are produced for the three different basis sets, then they provide a simple method of comparing the approximation given by the three types of series. However, as N approaches N_{max}, this technique underestimates the exact value because of the terms that are neglected in the summations for $k > N_{max}$.

Clearly, the $L^2(N)$ error measures the expected error obtained on truncation of the series in (4.82), (4.83), (4.86) at the $k = N$ term. However, because this is only an error measured in the L^2 norm of (4.94)–(4.96), it does not give a reliable estimate for the point-wise maximum error to be expected for the respective error curves $E_N(t)$. Although graphs of log_{10} of $L_P^2(N)$, $L_T^2(N)$, $L_F^2(N)$, versus N, clearly show the type of convergence obtained, they do not indicate what the point-wise error could be. The latter must be determined by inspection as there is no simple relationship between the maximum error in $E_N(t)$ on the interval and the $L^2(N)$ error at the same value of N. The maximum error in $E_N(t)$ is related to the uniform norm, or L^∞ norm, that is not easy to estimate and for this reason the L^2 norm is usually applied instead.

EXAMPLES 4.14 and 4.15

4.9 Applications

To demonstrate the utility of the concepts discussed in this chapter, applications are shown in EXAMPLES 2.17, 2.18, 2.21- 23, 4.14 to 4.20. These differ in the character of the function being approximated. One function is exponential in argument, another oscillatory, and the third a combination of the two. Therefore these applications are sufficiently general to be considered typical for a functional form $f(\lambda, x)$ where, without risk of confusion, in place of t, x is the real variable defined on a finite interval and $\lambda \in \mathbb{R}$, the space of real numbers, is either a discrete or continuous parameter on which the function also depends. The applications are sufficient to demonstrate how exponential convergence of the series may be detected and a suitable truncation N determined for a prerequisite error ϵ.

In the first application for the exponential function, consider the special case of $f(z, x) = e^{zx}$ with $z \in [0, Z]$, a continuous parameter. The Legendre or Chebyshev expansion coefficients of (2.52,2.54) are obtainable from the integrals

$$I_k^P(z) = \int_{-1}^{+1} e^{zx} P_k(x) dx, \qquad (4.102)$$

$$I_k^T(z) = \frac{2}{\pi} \int_{-1}^{+1} e^{zx} T(x) \frac{dx}{\sqrt{1-x^2}}, \qquad (4.103)$$

that are both computable from the analytical results

$$I_k^P(z) = 2 \sum_{i=0}^{\infty} \frac{j_{k+i}(z)}{i!} z^i, \qquad (4.104)$$

$$I_k^T(z) = 2 \sum_{i=0}^{\infty} \frac{J_{k+i}(z)}{i!} z^i, \qquad (4.105)$$

4.9 Applications

where j_m is the spherical Bessel function and J_m is the Bessel function of integer order, as defined in Sect. 2.6. The analytical result of (4.105) presented here is in addition to the expression in (2.59) which is simpler to compute. The result of (4.105) requires values of $J_m(z)$ and may be computed by recurrence and/or from the Chebyshev series in [9] from the Fortran code JLRCHB. In the examples, results for the Chebyshev case are shown in computations for coefficients in (4.104), (4.105) and the convolution product (2.159)–(2.161). The corresponding $L_T^2(N)$ errors are also shown and a comparison of the $L_T^2(N)$ error for $z = 1, 2, 4$, is shown in EXAMPLE 4.20, where the increasing complexity of the function requires a larger truncation index N for the same choice of ϵ in the L^2 norm of E^*.

When a trigonometric series is applied to e^{zx} the Fourier coefficients of (2.73), (2.74) are obtained on integrating by parts twice in

$$c_k = \int_{-\pi}^{+\pi} e^{zx} \cos(k\pi x) dx, \quad k = 0, 1, 2, 3, \ldots, \qquad (4.106)$$

and

$$s_k = \int_{-\pi}^{+\pi} e^{zx} \sin(k\pi x) dx, \quad k = 1, 2, 3, \ldots, \qquad (4.107)$$

to obtain

$$c_k = (-1)^k \left(\frac{e^z - e^{-z}}{z + z^{-2}(k\pi)^2} \right), \quad k = 0, 1, 2, 3, \ldots, \qquad (4.108)$$

and

$$s_k = -\frac{k\pi}{z} c_k, \quad k = 1, 2, 3, \ldots. \qquad (4.109)$$

For the special case with $z = 1$, the last two expressions are those previously derived in (2.178), (2.179)

$$c_k = (-1)^k \left(\frac{e - e^{-1}}{1 + (k\pi)^2} \right), \quad k = 0, 1, 2, 3, \ldots, \qquad (4.110)$$

and

$$s_k = -k\pi c_k, \quad k = 1, 2, 3, \ldots. \qquad (4.111)$$

The second application is for the function $f(m, x) = \sin(m\pi x)$ with $\lambda = m$, an integer. The Legendre and Chebyshev expansion coefficients may be obtained by numerical quadrature (see Chap. 7) and in the Chebyshev series it suffices to compute the $m = 1$ case and obtain Chebyshev coefficients for larger m from the convolution product results for (2.155)–(2.161) while exercising care with regard to the effects of truncation error in the summations.

The third and final application is for the spherical Bessel function, $j_m(r)$ that is defined on the positive interval $r \in [0, R]$. The Chebyshev series is in even polynomials and fortunately the coefficients satisfy a four term recurrence that is also stable, therefore they are easily computed (see Sect. 2.6 and EXAMPLES 2.21, 2.22).

EXAMPLES 4.16–4.20

4.10 Exercises

1. For the Chebyshev series of (2.110) of the spherical Bessel function $j_\ell(ax)$ use the coefficients $B_n^\ell(a)$ on the interval $x \in [0, 1]$ from EXAMPLE 2.21 to show

   (a) the L^2 error as n increases for order $\ell = 0$,
   (b) the value of the series at $x = 0.25, 0.5, 0.75$ for $n = 0, 1, 3$, using the explicit forms of the Chebyshev polynomials,
   (c) the error when the series values are compared to the exact value (2.104).

2. For the Chebyshev series of (2.110) for the spherical Bessel function use the coefficients $B_n^\ell(a)$ on the interval $x \in [0, 1]$ from EXAMPLE 2.22 to show

   (a) the L^2 error as n increases for order $\ell = 1$,
   (b) the value of the series at $x = 0.25, 0.5, 0.75$ for $n = 0, 1, 3$, using the explicit forms of the Chebyshev polynomials,
   (c) the error when the series values are compared to the exact value (2.105).

3. For Exercises 2 and 3 compute the asymptotic index of convergence (2.187).

4.11 Programming Problems

1. Write a code to compute the spherical Bessel function $j_\ell(r)$ on the interval $x \in [0, 1]$, $r = ax$, in a Chebyshev series such as (2.110) using the algorithm of EXAMPLES 2.21 to 2.23.
2. Expand this code to compute the Legendre series coefficients for the expansion of e^{zx} with $x \in [0, 1]$, $z = 1, 2$. For the series (2.51) use the expression (4.104).
3. Use the code to repeat an analysis similar to that of the Chebyshev series in EXAMPLES 4.15, 4.17 and compare convergence of the Legendre and Chebyshev series.
4. Compare the absolute errors of (4.99), (4.100) for the two series at selected values of the arguments $x \in [0, 1]$, $z = 1, 2$.

References

1. Kantorovich LV, Akilov GP (1982) Functional analysis, 2nd edn. Pergamon Press, New York, NY
2. Ljusternik LA, Sobolow WI (1979) Elemente der Funktionalanalysis. Verlag Harri Deutsch, Thun, Frankfurth/Main
3. Young N (1988) An introduction to Hilbert space. Cambridge University Press, Cambridge, UK

4. Milne RD (1980) Applied functional analysis an introductory treatment. Pitman Publishing Ltd., London, UK
5. Riesz F, -Nagy BS (1955) Functional analysis. Frederick Ungar Publishing Co., New York, NY
6. Glazman NI, Glazman IM (1981) Theory of linear operators in Hilbert space, vol I and II. Pitman Publishing Ltd., London, UK
7. Natanson IP (1965) Constructive function theory, vols I and II. Frederick Ungar Publishing Co., New York, NY
8. Achieser NI (1967) Vorlesungen über Approximationstheorie. Akademie-Verlag, Berlin
9. Delic G (1979) Chebyshev series for the spherical Bessel function JL(r). Comput Phys Commun 18:73–86

Operator Equations and Notation 5

5.1 Operator Equations

One objective of this book is the study of numerical approximation of operator equations of the generic type

$$Ax = Gy \qquad (5.1)$$

where $A, G \in L(\mathbf{B}, \mathbf{B})$ are operators on elements x, y of an arbitrary Banach space $\mathbf{B}$ and L is the space of all operators mapping $\mathbf{B}$ into itself. A Banach space is a complete normed space (see §21 of [1] and for the definition of completeness see §13). The nature of the sets $\{x\}$, or $\{y\}$ and the space to which they belong determine the character of the operators A, G. This statement is a rather general one and special cases are to be found in the following sections. For the purposes of this discussion examples of such spaces are $\mathbf{E}$, where the elements are vectors and A, G are matrices, or $\mathbf{C}[a, b]$, the space of continuous functions $x(t), y(t)$, defined on the interval $t \in [a, b]$, as discussed in the previous chapter. In the latter case typical operators could be differential operators

$$Ax = \sum_n \frac{d^n}{dt^n} x(t), \qquad (5.2)$$

where $n = 1, 2, \cdots$ denotes the order of the derivative, or integral operators

$$Gy = \int_a^t k(t, s) y(s) ds, \qquad (5.3)$$

Supplementary Information The online version contains supplementary material available at https://doi.org/10.1007/978-3-031-90178-2_5

© The Author(s), under exclusive license to Springer Nature Switzerland AG 2026
G. Delic, *Guide to Numerical Algorithm Design and Development*,
Texts in Computer Science, https://doi.org/10.1007/978-3-031-90178-2_5

or linear combinations of differential and integral operators with appropriate boundary conditions on $x(a)$, $x(b)$, $y(a)$, and $y(t)$. The case when the combination of operators is linear has been extensively studied over several centuries and solution methods for the corresponding operator equations are well developed for many special cases. The non-linear case, which may contain terms such as $x(t)\frac{dx(t)}{dt}$ or $[x(t)]^2$ (to show the simplest possibilities), is considerably more difficult to solve.

A special interesting case of (5.1) occurs if G is a scalar $\lambda \in \mathbb{R}$, the space of real numbers and is either a discrete or continuous parameter and A is a second-order differential operator as in

$$\left[\frac{d^2}{dt^2} - k^2\right] x(t) = \lambda x(t), \tag{5.4}$$

or, if $A = I$, the identity operator, $Ix = x$ and G is an integral operator

$$x(t) = \lambda \int_a^b k(t,s) y(s) ds. \tag{5.5}$$

Both (5.4) and (5.5) are examples of eigenvalue problems where the scalar, λ, has a spectrum of values that is either only discrete, or has both discrete and continuous branches. The specific behavior of the spectrum in the eigenvalue problem depends on the choice of boundary conditions and the exact form of the operator. Some examples are considered in Chap. 9, and a special case of discrete complex eigenvalues has been reported in [2].

A numerical approximation replaces the operator equation (5.1) by a truncated representation. The idea of this approximation is best viewed as a "projection" of the original solution x, defined over the whole space $\mathbf{B}$, onto an m-dimensional subspace $\mathbf{B}^m \subset \mathbf{B}$, through the action of a projection operator, $P_m x = x_m$, $P_m y = y_m$. Projection operators were described in more detail in Chapter 4. They produce components of the element x along direction coordinates in the subspace $\mathbf{B}^m$ and may be either orthogonal or oblique. For the operators A, G approximations A_m, G_m are likewise obtained, so that a truncated approximation to (5.1) is

$$\begin{aligned} x_m &\approx x, \\ y_m &\approx y, \\ A_m &\approx A, \\ G_m &\approx G. \end{aligned} \tag{5.6}$$

Study of convergence behavior by truncation as $m \to \infty$ is the primary source of information on the relative success (or failure) of an approximation scheme for the operator equation (5.1). This constitutes a dominant theme in this book and its study uses aspects of applied functional analysis described in Chap. 4. The next two subsections confine attention to the familiar space $\mathbf{E}$ of denumerably infinite sequences of numbers (vectors) and corresponding operators on them (matrices). To demonstrate a simple operator equation, the example for this section is Problem 18 from [3].

EXAMPLE 5.1

5.2 Vectors

Consider the subspace $\mathbf{E}^m \subset \mathbf{E}$ of denumerably infinite sequences of scalars (real or complex) and define elements $x \in \mathbf{E}^m$ as the array

$$x = \begin{bmatrix} x_1 \\ x_2 \\ \vdots \\ x_m \end{bmatrix}, \tag{5.7}$$

where the elements of the set $\{x_i\}_{i=1}^m$ are scalars. They could, also be operators, functions, or series of them, but in such a case this would not be the space $\mathbf{E}^m$. The parameter m denotes the length of the vector and determines the rank of the subspace. Computer solution of the generic operator equation (5.1) consists of reduction to the finite rank case and solution of the finite rank problem (5.6). The vector x may also be represented as a row, in which case it is said to be transposed

$$x^T = [x_1, x_2, \cdots, x_m]. \tag{5.8}$$

Characteristic arithmetic operations with vectors act on elements and are termed "elemental" as is shown in Table 5.1.

Observe that there are two types of product for vectors in Table 5.1. These correspond, respectively, to multiplication by a scalar and by a transposed vector. There is also the case of sums of scalar multiples of vectors known as a linear combination, where each component of the resultant vector is a sum of products as in

$$\sum_{j=1}^m \alpha_j x^{(j)} = \begin{bmatrix} \sum_{j=1}^m \alpha_j x_1^{(j)} \\ \sum_{j=1}^m \alpha_j x_2^{(j)} \\ \vdots \\ \sum_{j=1}^m \alpha_j x_m^{(j)} \end{bmatrix}. \tag{5.9}$$

When such a linear combination exists for a set of non-zero scalars $\{\alpha_j\}_1^m$, such that the resultant vector is zero, then the set of vectors, $\{x^{(j)}\}_1^m$, is a linearly dependent set.

Table 5.1 Vectors

Operation	Vector notation	Elemental definition
Equality	$x = y$	$x_i = y_i, i = 1, \ldots, m$
Addition	$x \pm y$	$x_i \pm y_i, i = 1, \ldots, m$
Scalar product	αx	$\alpha x_i, i = 1, \ldots, m$
Dot product	$x^T y$	$\sum_{i=1}^m x_i y_i$

Otherwise the set of vectors is linearly independent and no vector can be expressed as linear combination of the other vectors in the set.

EXAMPLE 5.2

5.3 Matrices

A matrix [4] is an array of objects arranged in rows and columns

$$A = \begin{bmatrix} a_{11} & a_{12} & \cdots & a_{1n} \\ a_{21} & a_{22} & \cdots & a_{2n} \\ \vdots & \vdots & \ddots & \vdots \\ a_{m1} & a_{m2} & \cdots & a_{mn} \end{bmatrix}, \tag{5.10}$$

where the coefficients a_{ij} may be scalars (real or complex), operators, functions, or series of them. The matrix rank may be finite or infinite but attention here is restricted to the finite rank case. In the example of (5.10) the matrix is said to be $m \times n$, and rectangular if there are m rows and n columns, but is square if $m = n$. Vectors are thus special cases of matrices as in (5.7,5.8) where x is $m \times 1$ and x^T is $1 \times m$. For simplicity this discussion is restricted to the square matrix case with $m = n$. The transpose, A^T, of a matrix A, is obtained by reflection about the main diagonal so that matrix element, a_{ij}^T, of the transposed matrix is the matrix element in the jth row and ith column of A. The matrix A is symmetric if $a_{ij} = a_{ji}$ and otherwise it is not symmetric. A hermitian matrix, $A^* = A$, is one where the complex conjugate of each matrix element of the transpose is the same as the original.

$$a_{ij}^T = a_{ji}, \tag{5.11}$$

$$a_{ij}^* = \bar{a}_{ij}^T, \tag{5.12}$$

$$= \bar{a}_{ji}. \tag{5.13}$$

These and other matrix properties are summarized in Table 5.2 where the elemental definition applies for $i, j = 1, \cdots, n$. A unitary matrix, or orthogonal matrix, has the additional property that $Q^{-1} = Q^* = Q^T$.

Matrices have structure described by the distribution of non-zero entries. Thus, a square diagonal matrix has non-zero entries only along the main diagonal

$$\Lambda = \begin{bmatrix} \lambda_{11} & 0 & \cdots & 0 \\ 0 & \lambda_{22} & & 0 \\ \vdots & & \ddots & \vdots \\ 0 & 0 & \cdots & \lambda_{nn} \end{bmatrix}, \tag{5.14}$$

5.3 Matrices

Table 5.2 Matrices

Operation	Vector notation	Elemental definition
Symmetry	$A^T = A$	$a_{ij}^T = a_{ji}$
Skew-symmetry	$A^T = -A$	$a_{ij}^T = -a_{ji}$
Hermiticity	$A^* = A$	$a_{ij}^* = \bar{a}_{ji}$
Skew-hermiticity	$A^* = -A$	$a_{ij}^* = -\bar{a}_{ji}$
Normality	$A^*A = AA^*$	$\sum_{k=1}^{n} a_{ik}^* a_{kj} = \sum_{k=1}^{n} a_{ik} a_{kj}^*$
Unitarity	$Q^*Q = I$	$\sum_{k=1}^{n} q_{ik}^* q_{kj} = 1$

and a special case of this is the identity matrix for which all diagonal entries are unity

$$I = \begin{bmatrix} 1 & 0 & \cdots & 0 \\ 0 & 1 & & 0 \\ \vdots & & \ddots & \vdots \\ 0 & 0 & \cdots & 1 \end{bmatrix}. \tag{5.15}$$

A banded matrix, A, has a bandwidth of $k+m+1$ when the only non-zero entries are those parallel to the main diagonal with $a_{ij} \neq 0, i-k \leq j \leq i+m$, for (non-negative) integers k, m. Other special matrix structures are summarized algebraically in the following matrix portraits. A tridiagonal matrix is defined by $a_{ij} = 0, |i-j| \geq 2$

$$A = \begin{bmatrix} a_{11} & a_{12} & 0 & \cdots & & 0 \\ a_{21} & a_{22} & a_{23} & & & \vdots \\ 0 & a_{32} & a_{33} & \ddots & & \vdots \\ \vdots & & \ddots & \ddots & \ddots & \vdots \\ \vdots & & & & a_{n-1n-1} & a_{n-1n} \\ 0 & \cdots & \cdots & \cdots & a_{nn-1} & a_{nn} \end{bmatrix}, \tag{5.16}$$

an upper triangular matrix has $a_{ij} = 0, i > j$,

$$A = \begin{bmatrix} a_{11} & a_{12} & a_{13} & \cdots & & a_{1n} \\ 0 & a_{22} & a_{23} & \cdots & & \vdots \\ \vdots & & a_{33} & \cdots & & \vdots \\ \vdots & & & \ddots & & \vdots \\ \vdots & & & & a_{n-1n-1} & a_{n-1n} \\ 0 & \cdots & \cdots & \cdots & 0 & a_{nn} \end{bmatrix}, \tag{5.17}$$

and an upper hessenberg matrix has $a_{ij} = 0, i > j+1$,

$$A = \begin{bmatrix} a_{11} & a_{12} & a_{13} & \cdots & & \cdots & a_{1n} \\ a_{21} & a_{22} & a_{23} & \cdots & & \cdots & \vdots \\ 0 & a_{32} & a_{33} & \cdots & & \cdots & \vdots \\ \vdots & \vdots & \ddots & & & & \vdots \\ \vdots & \vdots & & a_{n-1n-2} & a_{n-1n-1} & a_{n-1n} \\ 0 & \cdots & \cdots & & 0 & a_{nn-1} & a_{nn} \end{bmatrix}. \tag{5.18}$$

The corresponding lower triangular and hessenberg matrices have $a_{ij} = 0$, for $i < j$ and $i < j+1$, respectively. A matrix is diagonally dominant if

$$\mid a_{ii} \mid \geq \sum_{j \neq i} \mid a_{ij} \mid, i = 1, \cdots, n. \tag{5.19}$$

A sparse $n \times n$ matrix is one where the number of non-zero elements $\{a_{ij}\}$ is much smaller than the total number of elements in the matrix. Sparse matrices are not stored as full $n \times n$ matrices but in specialized storage schemes that usually require an auxiliary index array to store values of subscripts for locations of non-zero elements. The rank of the index matrix is determined by the number of non-zero elements and the sparse storage scheme chosen to represent the sparse matrix. See Sect. 5.4 for more detail.

Just as in the case of the vectors, the characteristic arithmetic matrix operations are on the matrix elements and Table 5.3 summarizes some typical operations.

Note that the resultant of the matrix-vector product is a vector while that of the matrix-matrix product is another matrix. For the matrix element product, the component matrices must have shapes that conform. Compare also the product $x^T y$ of Table 5.1 which has a scalar resultant to that of yx^T that is a matrix. Another simple application of results in Tables 5.1 and 5.2 is in the definition of a positive definite

Table 5.3 Matrix operations

Operation	Matrix notation	Elemental definition
Equality	$A = B$	$a_{ij} = b_{ij}, i, j = 1, \ldots, n$
Addition	$A \pm B$	$a_{ij} \pm b_{ij}, i, j = 1, \ldots, n$
Scalar product	αA	$\alpha a_{ij}, i, j = 1, \ldots, n$
Matrix-vector product	Ax	$\sum_{j=1}^{n} a_{ij} x_j, i = 1, \ldots, n$
Matrix-matrix product	AB	$\sum_{j=1}^{n} a_{ij} b_{jk}, i, k = 1, \ldots, n$
Matrix-element product	$A \times B$	$a_{ij} b_{ij}, i, j = 1, \ldots, n$
Trace	$Tr(A)$	$\sum_{j=1}^{n} a_{ii}$

5.3 Matrices

rank n matrix A for which

$$x^T A x = \sum_{i=1}^{n} \sum_{j=1}^{n} x_i a_{ij} x_j > 0. \qquad (5.20)$$

The result of this operation is a scalar that follows from the matrix-vector product of Table 5.3 and the dot product of Table 5.1. Further properties of matrices are as follows for sums

$$(A + B) + C = A + (B + C), \qquad (5.21)$$
$$(A + B)^T = A^T + B^T, \qquad (5.22)$$

and for matrix-matrix products

$$A(B + C) = AB + AC, \qquad (5.23)$$
$$(AB)C = A(BC), \qquad (5.24)$$
$$(AB)^T = B^T A^T. \qquad (5.25)$$

In general the matrix-matrix product of two matrices does not commute

$$AB \neq BA, \qquad (5.26)$$

otherwise A, B are commuting matrices. In particular if I is the identity matrix then

$$AI = IA, \qquad (5.27)$$

and a special case of the matrix product occurs if

$$AB = BA = I, \qquad (5.28)$$

when B is said to be the inverse of A, or A is the inverse of B, and this is written

$$A = B^{-1}, \qquad (5.29)$$
$$B = A^{-1}. \qquad (5.30)$$

For the inverse of a product, just as for the transpose of a product, we have the results

$$(AB)^{-1} = B^{-1} A^{-1}, \qquad (5.31)$$
$$(A^{-1})^T = (A^T)^{-1}. \qquad (5.32)$$

EXAMPLE 5.3

5.4 Sparse Matrix Storage

A sparse matrix [5,6] is one where the number of non-zero elements are fewer than the total number of elements of a matrix. If NZ is the number of non-zero elements in a sparse matrix A, of rank N, then, typically, $NZ << N^2$. All sparse matrix algorithms reference only non-zero elements and store the value and indices in arrays, but they differ in the storage method for the row and column location in the full matrix. A brief summary of the three storage methods are as follows.

- The Triplet Storage (TS) scheme scans rows and columns of the matrix A, then stores column and row index values in two integer arrays of length NZ, such that $k = 1, \cdots, NZ$. Matrix element indices and values are stored as follows:

$$\text{value}(k) = a_{ij}, \qquad (5.33)$$
$$\text{row}(k) = i, \qquad (5.34)$$
$$\text{col}(k) = j. \qquad (5.35)$$

- The Compressed Column (CC) storage scheme scans down successive columns and uses an integer array $i(k), k = 1, \cdots, NZ$, of length NZ, together with another pointer array $p(j), j = 1, \cdots, N+1$, of length $N+1$, so that row indices are stored in integer arrays $i(p(j))$ through $i(p(j+1)-1)$. The CC scheme is described in Chapter 2 of Davis [6] for the C language case (where index counters begin at 0).
- The Compressed Row (CR) storage scheme scans across successive rows and uses a similar pointer scheme as in the CC method above and is described in Chapter 6 of [7].

To demonstrate these storage schemes, consider a simple rank $N = 4$ sparse matrix [6] that has 16 elements, of which 10 are non-zero

$$A = \begin{bmatrix} a_{11} & 0 & a_{13} & 0 \\ a_{21} & a_{22} & 0 & a_{24} \\ 0 & a_{32} & a_{33} & 0 \\ a_{41} & a_{42} & 0 & a_{44} \end{bmatrix}. \qquad (5.36)$$

The triplet scheme then stores entries as follows with three vectors of length $NZ = 10$

$$\begin{array}{rcccccccccc} a_{ij} = & a_{11} & a_{13} & a_{21} & a_{22} & a_{24} & a_{32} & a_{33} & a_{41} & a_{42} & a_{44} \\ \text{row}(k) = & 1 & 1 & 2 & 2 & 2 & 3 & 3 & 4 & 4 & 4 \\ \text{col}(k) = & 1 & 3 & 1 & 2 & 4 & 2 & 3 & 1 & 2 & 4 \end{array} \qquad (5.37)$$

and, in matrix operations, this requires look-up in two arrays, row(k), col(k) to find the two indices before the matrix element value may be extracted. In computer code these are both indirect memory references to extract values of the matrix required for arithmetic operations. Indirect because they reference a subscipt of a subscript.

5.4 Sparse Matrix Storage

Table 5.4 CC operations

j	$p(j)$	$i(p(j)),\cdots,i(p(j+1)-1)$
1	1	1, 2, 4
2	4	2, 3, 4
3	7	1, 3
4	9	2, 4

In the CC scheme the matrix is stored as dense column vectors, where for each vector, non-zero elements are stored in compressed form and this has advantages in matrix operations (as is discussed in the case study of Sect. 5.12). The method still uses an integer array $i(k), k = 1, \cdots, NZ$, to store row indices. A value array $\alpha(k)$ has the same length, while a pointer array, $p(j), j = 1, \cdots, N+1$, is used to retrieve row indices for column j stored in $i(p(j))$ through $i(p(j+1)) - 1$. The numerical values are stored in the same locations in the value array. In the CC scheme the mapping is as follows for the simple example of (5.36)

$$\begin{aligned}
\alpha(k) &= a_{11}\ a_{21}\ a_{41}|\ a_{22}\ a_{32}\ a_{42}|\ a_{13}\ a_{33}|\ a_{24}\ a_{44}| \\
i(k) &= 1\ \ 2\ \ 4\ \ \ 2\ \ 3\ \ 4\ \ \ 1\ \ 3\ \ \ 2\ \ 4 \\
p(j) &= 1\ \ \ \ \ \ \ \ \ \ \ 4\ \ \ \ \ \ \ \ \ \ \ 7\ \ \ \ 9\ \ \ \ \ \ 11
\end{aligned} \qquad (5.38)$$

where, $p(N+1) = NZ + 1$, Table 5.4 shows how row indices (and values) are stored in this example.

The CR scheme is similar in approach to the CC method, but scans across rows as is described in [7].

Each sparse storage scheme requires indirect subscript references at some level and the implementation has consequences for parallel algorithm opportunities. For example, the TS scheme, when coded, uses complex loop ranges based on indirect array references that are evaluated only at runtime. This prohibits parallelization of outer loop nests because of the indirect addressing and is a major performance inhibitor on modern commodity processors that do not support hardware gather/scatter instructions. As a consequence, this leads to excessive translation look-aside buffer (TLB) cache look-up that inevitably stalls arithmetic processing on a commodity processor chip. This was not always the case with support of hardware gather/scatter operations on former architectures such as the Cray^{TM} vector/parallel computers and the Intel Phi processor Many-Integrated-Core^{TM} (MIC) architectures. However, these are now end-of-life computer resources and coding for current commodity processors requires care in implementation of sparse matrix-vector operations. An application of the CC storage scheme as a replacement of the TS scheme is described in [8] for the Community Multiscale Air Quality model (CMAQ) and the Fortran code FSPARSE example is a schematic version of the algorithm.

EXAMPLE 5.4 and FORTRAN code FSPARSE

5.5 Determinants

The determinant $|A|$ of a matrix A is a scalar quantity computed from the array which constitutes the matrix. For a rank 2 matrix the determinant of order 2 is defined as

$$|A| = \begin{vmatrix} a_{11} & a_{12} \\ a_{21} & a_{22} \end{vmatrix},$$

$$= a_{11}a_{22} - a_{12}a_{21}, \tag{5.39}$$

while for a rank 3 matrix the determinant of order 3 is expanded in determinants of order 2 as in

$$|A| = \begin{vmatrix} a_{11} & a_{12} & a_{13} \\ a_{21} & a_{22} & a_{23} \\ a_{31} & a_{32} & a_{33} \end{vmatrix},$$

$$= a_{11} \begin{vmatrix} a_{22} & a_{23} \\ a_{32} & a_{33} \end{vmatrix} - a_{12} \begin{vmatrix} a_{21} & a_{23} \\ a_{31} & a_{33} \end{vmatrix} + a_{13} \begin{vmatrix} a_{21} & a_{22} \\ a_{31} & a_{32} \end{vmatrix},$$

$$= a_{11}A_{11} + a_{12}A_{12} + a_{13}A_{13}. \tag{5.40}$$

The order 2 determinants of (5.40) are called minors M_{1j} and are computed by analogy to (5.39). Each rank $n-1$ minor is formed from the rank n determinant by deleting the row and column corresponding to the respective subscripts. When the sign is included with the minors then co-factors, such as those in (5.40), are defined by $A_{1j} = (-1)^{1+j} M_{1j}$ for the order 2 determinant. However, observe that the coefficients, a_{1j}, of the co-factors in (5.40) correspond to the first row of A. In general, there is no such restriction and a rank n matrix has an order n determinant defined by

$$|A| = \sum_{j=1}^{n} a_{ij} A_{ij}, \tag{5.41}$$

where the subscript i has the range $1 \leq i \leq n$. In this definition, subscripts i and j can be interchanged and the cofactors of (5.41) are defined by

$$A_{ij} = (-1)^{i+j} M_{ij}, \tag{5.42}$$

with minors M_{ij} as determinants of order $n-1$ obtained by deleting the row and column corresponding to the subscripts i and j, respectively.

An interesting special case occurs when the matrix is upper triangular. Consider the rank three matrix for which (5.40) gives the result

$$|A| = \begin{vmatrix} a_{11} & a_{12} & a_{13} \\ 0 & a_{22} & a_{23} \\ 0 & 0 & a_{33} \end{vmatrix}, \tag{5.43}$$

$$=a_{11}\begin{vmatrix} a_{22} & a_{23} \\ 0 & a_{33} \end{vmatrix} - a_{12}\begin{vmatrix} 0 & a_{23} \\ 0 & a_{33} \end{vmatrix} + a_{13}\begin{vmatrix} 0 & a_{22} \\ 0 & 0 \end{vmatrix}, \qquad (5.44)$$

$$=a_{11}a_{22}a_{33} - a_{12} \times 0 + a_{13} \times 0, \qquad (5.45)$$

and the determinant is simply the product of the diagonal elements of the upper triangular matrix. In the rank n case the determinant of order n is the result of the product of all n diagonal matrix elements of a triangular matrix. Some further properties of the determinant $\mid A \mid$ of a matrix A are as follows:

1. Interchange of a pair of rows or columns changes the sign of $\mid A \mid$.
2. When two rows or columns are identical then $\mid A \mid = 0$.
3. If all elements of a row or column are multiplied by the same constant then so is the determinant.
4. If a row or column has only zero entries then $\mid A \mid = 0$.
5. If a scalar multiple of the elements of any row (or column) is added to the corresponding elements of another row (or column) then the determinant is unchanged.
6. For the transposed matrix it follows that $\mid A^T \mid = \mid A \mid$.
7. If B is a matrix then for corresponding determinants $\mid AB \mid = \mid A \mid \times \mid B \mid$.
8. Two conditions are possible: either $\mid A \mid = 0$, and matrix A is said to be singular, or $\mid A \mid \neq 0$ and matrix A is not singular.

Another property of an $n \times n$ matrix A is the permanent, $per(A)$, defined as the sum of all products of the form

$$a_{1,b_1} a_{2,b_2} a_{3,b_3} \cdots a_{n-1,b_{n-1}} a_{n,b_n}, \qquad (5.46)$$

where the n-set $\{b_1 b_2 b_3 \cdots b_{n-1} b_n\}$ is the set of n! permutations of the index $j = 1, \ldots, n$. The value of the permanent is an invariant under row/column permutations of A. Unlike the determinant, the permanent of a matrix has no alternating sign for separate terms and lacks simple properties such as $per(AB) = per(A)per(B)$. In the simplest algorithm computation of $per(A)$ devolves into three disjoint parts:

1. generation of the n-set $\{b_1 b_2 b_3 \cdots b_{n-1} b_n\}$,
2. computation of the product (5.46) for each term of the n-set, and
3. summation of the results produced in step (2).

For example, with $n = 4$ there are 24 terms in the sum. However, for large n the values of $n!$ make this simple algorithm prohibitively expensive in computation time. In the case of 0, 1 matrices (i.e., those with matrix elements having values of either 0 or 1) the value of $per(A)$ is simply the size of the minimal n-set because non-zero terms of (5.46) are unity. Special cases where the value of the permanent is known include

$$a_{i,j} = 1, \; i, j = 1, \ldots, n,$$
$$per(A) = n! \tag{5.47}$$

and

$$a_{i,j} = 1, \; i, j = 1, \ldots, n, \; i \neq j,$$
$$= 0, \; i = j,$$
$$per(A) = n! \left[\frac{1}{2!} - \frac{1}{3!} + \cdots + \frac{(-1)^n}{n!} \right]. \tag{5.48}$$

The permanent has a long history and, because of its importance in combinatorial mathematics, continues to be investigated. At least four algorithms for computing the permanent (three including source code) have been published in the last 18 years. For further details, see the citations in [9] where a specific algorithm for (0,1) matrices designed by the author has been published.

EXAMPLES 5.5 and 5.6
EXAMPLE Fortran code PERM

5.6 Matrix Operator Equations

A simple example of the operator equation (5.1) is the set of three linear equations

$$a_{11}x_1 + a_{12}x_2 + a_{13}x_3 = y_1,$$
$$a_{21}x_1 + a_{22}x_2 + a_{23}x_3 = y_2,$$
$$a_{31}x_1 + a_{32}x_2 + a_{33}x_3 = y_3, \tag{5.49}$$

which may also be written in matrix-vector notation

$$\begin{bmatrix} a_{11} & a_{12} & a_{13} \\ a_{21} & a_{22} & a_{23} \\ a_{31} & a_{32} & a_{33} \end{bmatrix} \begin{bmatrix} x_1 \\ x_2 \\ x_3 \end{bmatrix} = \begin{bmatrix} y_1 \\ y_2 \\ y_3 \end{bmatrix}, \tag{5.50}$$

or, in the form of (5.1) $Ax = y$. In (5.50) the vector x is the unknown while both A and y are known data. One method of solution for x is Cramer's rule which gives each unknown as the ratio of two determinants. The denominator in this ratio is $|A|$, but the numerator is the determinant obtained when the column of coefficients of the corresponding unknown is replaced by the vector y. As an example, in the case of x_1,

$$x_1 = \frac{1}{|A|} \begin{vmatrix} y_1 & a_{12} & a_{13} \\ y_2 & a_{22} & a_{23} \\ y_3 & a_{32} & a_{33} \end{vmatrix}, \tag{5.51}$$

and similarly for the other two unknowns.

5.6 Matrix Operator Equations

Matrix systems such as those in (5.50) have a unique solution for each choice of the vector y if the determinant, $|A|$, of A is non-zero, or equivalently, the matrix A is nonsingular. This always is the case for a positive definite matrix A that satisfies (5.20) and may be demonstrated by assuming that $Ax = 0$ has a solution $x \neq 0$. The latter implies that $x^T A x = 0$ which contradicts the assumption that A is positive definite. Thus only $x = 0$ is a solution of $Ax = 0$. A unique solution to $Ax = y$ exists for each y only if A is also nonsingular. Therefore a positive definite matrix A is also nonsingular.

Another method of determining the unknowns in $Ax = y$ is to compute the inverse of A defined by (5.29,5.30). The importance of the inverse A^{-1} is that it solves the operator equation $Ax = y$. This may be shown by left multiplication by A^{-1}

$$A^{-1}Ax = A^{-1}y, \tag{5.52}$$

and, because $A^{-1}Ax = Ix = x$, it follows that

$$x = A^{-1}y. \tag{5.53}$$

Another matrix operator is the analogue of (5.4) and (5.5) for the eigenvalue problem

$$Ax^{(i)} = \lambda_i x^{(i)},$$
$$AX = X\Lambda, \tag{5.54}$$

where the last result uses matrix notation with X as the matrix whose columns are the eigenvectors $x^{(i)}$, $i = 1, \cdots, n$, and Λ is the diagonal matrix of eigenvalues. In general, eigenvectors $x^{(i)}$ have the property of orthogonality defined by the inner product $(\cdot,\cdot)$ (see Sect. 4.6) for a matrix A of rank n

$$\left(x^{(i)}, x^{(k)}\right) = \sum_{j=1}^{n} x_j^{(i)} x_j^{(k)},$$
$$= 0, \ i \neq k,$$
$$\neq 0, \ i = k. \tag{5.55}$$

If the eigenvectors are normalized, or scaled, so that the non-zero constant in (5.55) for $i = k$ is unity, then the eigenvectors are said to be orthonormal and

$$X^T X = X X^T = I. \tag{5.56}$$

It then follows from (5.54,5.56) and the property $\Lambda X = X \Lambda$ that

$$A = X \Lambda X^T, \tag{5.57}$$
$$X^T A X = \Lambda, \tag{5.58}$$

and the matrix operator is said to have been "diagonalized", or expressed in diagonal form. The eigenvalues λ can be real or complex. If $\bar{\lambda}$ denotes the complex conjugate of λ, then whenever matrix A has an eigenvalue λ, A^* has an eigenvalue $\bar{\lambda}$. The corresponding eigenvectors are distinguished as right (x) and left (z) eigenvectors, respectively, in

$$Ax = \Lambda x, \qquad (5.59)$$

$$z^*A = \Lambda z^*, \qquad (5.60)$$

$$x^*A^* = \bar{\Lambda} x^*, \qquad (5.61)$$

$$A^*z = \bar{\Lambda} z. \qquad (5.62)$$

The general matrix eigenvalue problem has been studied by the author in [2] where complex eigenvalues may occur.

When no distinction is made, the right eigenvector is assumed. It is often important to know if a set of (not necessarily orthonormal) vectors, $x^{(i)}$, is linearly independent. That is, no set of non-zero constants $\{c_i\}_1^n$ exists such that

$$c_1 x^{(1)} + c_2 x^{(2)} + \cdots + c_n x^{(n)} = 0. \qquad (5.63)$$

One way to proceed with this test is to use the inner product in (5.55) and form the Gram matrix, or Gramian

$$G = \begin{bmatrix} (x^{(1)}, x^{(1)}) & (x^{(1)}, x^{(2)}) & \cdots & (x^{(1)}, x^{(n)}) \\ (x^{(2)}, x^{(1)}) & (x^{(2)}, x^{(2)}) & & (x^{(2)}, x^{(n)}) \\ \vdots & \vdots & \ddots & \vdots \\ (x^{(n)}, x^{(1)}) & (x^{(n)}, x^{(2)}) & & (x^{(n)}, x^{(n)}) \end{bmatrix}, \qquad (5.64)$$

then, for the existence of a non-zero set $\{c_i\}_1^n$ in (5.63), the determinant of the Gramian matrix $|G|$ must be zero (see Chapter IX §3 of [4]). Conversely, the set of vectors $\{x^{(i)}\}_1^n$ is linearly independent if $|G| \neq 0$.

The solution of linear systems of equations is the subject of the next section, where some standard numerical methods of obtaining the inverse and solving matrix operator equations is investigated in more detail. The treatment is a basic one and the reader is encouraged to view more advanced treatments of linear equation solution methods such as those in [5–7, 10–34].

EXAMPLE 5.7

5.7 Matrix Solution Methods

This section explores some basic numerical approaches to solving a system of linear equations in the form (5.50), here written as a rank n system of equations in matrix-vector form

5.7 Matrix Solution Methods

$$\begin{bmatrix} a_{11} & a_{12} & \cdots & a_{1n} \\ a_{21} & a_{22} & \cdots & a_{2n} \\ \vdots & \vdots & \ddots & \vdots \\ a_{n1} & a_{n2} & \cdots & a_{nn} \end{bmatrix} \begin{bmatrix} x_1 \\ x_2 \\ \vdots \\ x_n \end{bmatrix} = \begin{bmatrix} y_1 \\ y_2 \\ \vdots \\ y_n \end{bmatrix}. \qquad (5.65)$$

The objective here is only to describe some typical numerical solution methods and outline inherent problems. The discussion is rudimentary and in no way comes close to satisfying the broad scope of this subject. The methods discussed here should not be considered as the best available for applications in high level model code. Instead, once the concepts have been understood, the reader should seek out the many excellent subroutines or function libraries that have stable numerical algorithms for linear systems such as (5.65). Such program libraries avoid many of the pitfalls encountered in solving large systems of linear equations (see Sect. 5.14 for examples). Furthermore, many of these libraries are now available with code optimized for vector and/or parallel performance and offer orders of magnitude better performance than the scalar algorithms discussed here. Therefore, the discussion of this chapter is pedagogical in character and aims to shed light on how easily problems arise in numerical methods for linear systems.

Two broad classes of numerical solution methods for linear systems of equations in common use are discussed here. Each approach has its merits and applicability is related to the specific character of the linear equations whose solution is required. The first class of solution method is that of direct algebraic solution such as Gaussian and Gauss-Jordan elimination discussed in Sects. 5.7.1 to 5.7.7. The second class uses iterative techniques to solve for an approximate solution and simple examples are the Jacobi and Gauss-Seidel iteration discussed in Sect. 5.9. In this respect the concepts developed in Chap. 4 find a direct application in the measure of convergence properties. Convergence criteria and relaxation, or convergence acceleration, are discussed in Sects. 5.10 and 5.11, respectively. Understanding of convergence properties is enhanced by application of norm properties developed in Chap. 4 and this is applied to matrices in Sect. 5.8 where stability conditions are defined. For the discussion in the following sections it is assumed that the basics of vectors, matrices and determinants, developed above, have been assimilated and that the general discussion of normed spaces in Chap. 4, have been studied.

5.7.1 Gaussian Elimination

In Gaussian elimination successive pairs of equations are utilized to eliminate each unknown in turn until the last equation is reached with only one unknown. Then, the resultant system is used in a back substitution process to determine all unknowns, starting with the last one. To demonstrate the procedure consider a simple rank 3 system, as in (5.49), with the assumption that $a_{11} \neq 0$. The unknown x_1 is eliminated by forming the multipliers

$$m_{21} = \frac{a_{21}}{a_{11}}, \tag{5.66}$$

$$m_{31} = \frac{a_{31}}{a_{11}}. \tag{5.67}$$

These multipliers are applied to the second and third rows of (5.49) in turn and are then subtracted from each, respectively, to yield a modified set of equations, where the first row is unchanged

$$\begin{aligned} a_{11}x_1 + a_{12}x_2 + a_{13}x_3 &= y_1, \\ a_{22}^{<2>} x_2 + a_{23}^{<2>} x_3 &= y_2^{<2>}, \\ a_{32}^{<2>} x_2 + a_{33}^{<2>} x_3 &= y_3^{<2>}. \end{aligned} \tag{5.68}$$

The superscript on the coefficients indicates the version number of the corresponding matrix elements of A. The modified terms are the result of the operations

$$a_{ij}^{<2>} = a_{ij} - m_{i1}a_{1j}, \quad i, j = 2, 3, \tag{5.69}$$

$$y_i^{<2>} = y_i - m_{i1}y_1, \quad i, = 2, 3. \tag{5.70}$$

Assuming that $a_{22}^{<2>} \neq 0$, x_2 is eliminated in the third row by forming a new multiplier

$$m_{32} = \frac{a_{32}^{<2>}}{a_{22}^{<2>}}, \tag{5.71}$$

to give a second set of modified equations, where only the third row is changed

$$\begin{aligned} a_{11}x_1 + a_{12}x_2 + a_{13}x_3 &= y_1, \\ a_{22}^{<2>} x_2 + a_{23}^{<2>} x_3 &= y_2^{<2>}, \\ a_{33}^{<3>} x_3 &= y_3^{<3>}, \end{aligned} \tag{5.72}$$

where

$$a_{33}^{<3>} = a_{33}^{<2>} - m_{32}a_{23}^{<2>}, \tag{5.73}$$

$$y_3^{<3>} = y_3^{<2>} - m_{32}y_2^{<2>}. \tag{5.74}$$

The result of (5.72), when represented in matrix form, is

$$A^{<3>}x = y^{<3>}, \tag{5.75}$$

and the matrix, $A^{<3>}$, has the upper triangular structure of (5.17), with the solution vector unchanged, while the vector, y, has been modified. Thus the solution of (5.75) is also the solution of the unmodified linear system. This solution is easily obtained from (5.75) by a process of back substitution beginning in the last equation with

5.7 Matrix Solution Methods

$$x_3 = \frac{y_3^{<3>}}{a_{33}^{<3>}}, \tag{5.76}$$

and for the second and first rows of (5.72), respectively

$$x_2 = \frac{y_2^{<2>} - a_{23}^{<2>} x_3}{a_{22}^{<2>}}, \tag{5.77}$$

$$x_1 = \frac{y_1 - a_{12} x_2 - a_{13} x_3}{a_{11}}. \tag{5.78}$$

Thus Gaussian elimination is seen to be a two step process, the first is elimination of unknowns and the second is back substitution for sequential solution of the unknowns. If the system is of rank n, then there are $n-1$ elimination steps and n back substitution steps. If $y_i^{<1>} = y_i$, and $a_{ij}^{<1>} = a_{ij}$, then, at step k, in this general case, unknowns x_1 to x_{k-1} are eliminated and the linear system at the kth step is reduced to the form

$$\begin{aligned}
a_{11}^{<1>} x_1 + a_{12}^{<1>} x_2 + \cdots + a_{1k}^{<1>} x_k + \cdots + a_{1n}^{<1>} x_n &= y_1^{<1>} \\
a_{22}^{<2>} x_2 + \cdots + a_{2k}^{<2>} x_k + \cdots + a_{2n}^{<2>} x_n &= y_2^{<2>} \\
&\vdots \\
a_{kk}^{<k>} x_k + \cdots + a_{kn}^{<k>} x_n &= y_k^{<k>} \\
&\vdots \\
a_{nk}^{<k>} x_k + \cdots + a_{nn}^{<k>} x_n &= y_n^{<k>}.
\end{aligned} \tag{5.79}$$

At the kth step, with the assumption, $a_{kk}^{<k>} \neq 0$, multipliers are formed from

$$m_{ik} = \frac{a_{ik}^{<k>}}{a_{kk}^{<k>}}, \quad i = k+1, \cdots, n, \tag{5.80}$$

and x_k is eliminated for rows $k+1$ to n by application of

$$a_{ij}^{<k+1>} = a_{ij}^{<k>} - m_{ik} a_{kj}^{<k>}, \quad i, j = k+1, \cdots, n, \tag{5.81}$$

$$y_i^{<k+1>} = y_i^{<k>} - m_{ik} y_k^{<k>}, \quad i = k+1, \cdots, n. \tag{5.82}$$

Gaussian elimination does not affect the vector x, but modifies the matrix A, through a sequence of transformations

$$A^{<1>} \to A^{<2>} \to \cdots \to A^{<n-1>} \to A^{<n>}, \tag{5.83}$$

where the limit of the sequence, $A^{<n>}$, is an upper triangular matrix. Once the transformations have been completed, then back substitution in the rank n case is a straightforward generalization of (5.75) to (5.78).

$$x_n = \frac{y_n^{<n>}}{a_{nn}^{<n>}}, \tag{5.84}$$

$$x_i = \frac{y_i^{<i>} - \sum_{j=i+1}^{n} a_{ij}^{<i>} x_j}{a_{ii}^{<i>}}, \quad i = n-1, \cdots, 1. \tag{5.85}$$

Clearly, if n is large, then a large number of arithmetic operations are required for both forward elimination and backward substitution. A detailed estimate of the number of addition, subtraction and multiplication operations for elimination shows that they increase as $\mathcal{O}(n^3)$, while those for back substitution increase as $\mathcal{O}(n^2)$. Therefore, the most expensive part of the calculation is always the forward elimination and this increases drastically with n. However, once the expensive part of the computation, namely, forward elimination has been completed, then back substitution may be applied repeatedly in those cases where the matrix is unchanged and different vectors y are applied.

EXAMPLES 5.8 and 5.9

5.7.2 Partial Pivoting

If the denominator $a_{kk}^{<k>}$ in the multiplier (5.80) is zero or small, relative to the numerator, then Gaussian elimination fails. Consider a simple rank 2 example to demonstrate how this occurs

$$\begin{aligned} \alpha x_1 + x_2 &= 1, \\ x_1 + x_2 &= 3. \end{aligned} \tag{5.86}$$

This has the solution $x_1 = 2$, $x_2 = 1$, if $\alpha = 0$. Now apply Gaussian elimination by computing the multiplier of (5.66) as

$$m_{21} = \alpha^{-1}, \tag{5.87}$$

for which the modified equations become

$$\begin{aligned} \alpha x_1 + x_2 &= 1, \\ (1 - \alpha^{-1}) x_2 &= 3 - \alpha^{-1}, \end{aligned} \tag{5.88}$$

and, from back substitution

$$x_2 = \frac{3 - \alpha^{-1}}{1 - \alpha^{-1}}, \tag{5.89}$$

$$x_1 = \frac{1 - x_2}{\alpha}. \tag{5.90}$$

If α is zero then the value of x_1 is infinite. However, if α is not zero, but still small, then the resulting solution has $x_2 \approx 1$, $x_1 \approx 0$ because α^{-1} is a large number.

Clearly, this is not the correct solution. The problem arises if α is small relative to the other coefficients and is not resolved by dividing (5.89) by α. However, consider the case of (5.86) in reverse order

$$x_1 + x_2 = 3,$$
$$\alpha x_1 + x_2 = 1, \tag{5.91}$$

in this case the multiplier (5.66) is

$$m_{21} = \alpha, \tag{5.92}$$

for which the modified equations are

$$x_1 + x_2 = 3,$$
$$(1 - \alpha)x_2 = 1 - 3\alpha, \tag{5.93}$$

and from back substitution

$$x_2 = \frac{3 - \alpha}{1 - \alpha}, \tag{5.94}$$
$$x_1 = 3 - x_2. \tag{5.95}$$

This results in the solution $x_2 \approx 1$, $x_1 \approx 2$, which approximates the exact solution well, if α is a small number. In changing the equation order, the modified Gaussian elimination method moves the smaller coefficient $a_{kk}^{<k>}$ further down and thereby ensures that the denominator in the multiplier is as far away from the zero value as possible.

The method described above is Gaussian elimination with partial pivoting. This seeks to avoid a catastrophic failure at the kth step by seeking the coefficient of largest magnitude in the remaining equations

$$\max_{k < i < n} |a_{ik}^{<k>}|, \tag{5.96}$$

and then, using the equation that this corresponds to as the pivot for the next elimination step. Partial pivoting swaps the order of the two equations and thereby increases the probability that

$$m_{ik} \leq 1, \ 1 \leq k < i \leq n, \tag{5.97}$$

and this reduces round-off error propagation. This pivot method is only partial because it interchanges rows whereas full pivoting operates on columns as well and linear algebra libraries include options for partial pivoting.

EXAMPLES 5.10 and 5.11

5.7.3 The Inverse Matrix

As one example of an application of Gaussian elimination we consider the numerical computation of the inverse matrix. Recall the discussion in Sect. 5.3 and

(5.29,5.30,5.53) that defined the inverse. In view of these results and property 7 in Sect. 5.5 for the product of two determinants, a matrix will only have an inverse if its determinant is not zero. Furthermore, if a matrix A has an inverse, then it is unique. This may be shown by the following argument for a square matrix. Assume matrix A has two inverses B_1, B_2, then

$$AB_1 = B_1 A = I, \qquad (5.98)$$
$$AB_2 = B_2 A = I, \qquad (5.99)$$

and left multiplication of (5.99) by B_1 yields

$$B_1 A B_2 = (B_1 A) B_2 \qquad (5.100)$$
$$= B_1 (A B_2) \qquad (5.101)$$
$$= I B_2 \qquad (5.102)$$
$$= B_1 I, \qquad (5.103)$$

but from (5.27) it follows that $B_1 = B_2$ which contradicts the assumption. The converse may also be shown and as a consequence the inverse of a matrix is unique.

The importance of the inverse is in the fact that it solves the linear system $Ax = y$ by $x = A^{-1} y$. While the linear system may simply be solved by Gaussian elimination, it is instructive to also show how it may be solved from (5.53) by first computing the inverse. The inverse may be obtained by solving for the matrix X in

$$AX = I, \qquad (5.104)$$

where the identity matrix has unity along the diagonal and zeros for all other entries. The rank n identity matrix may therefore be defined as a matrix whose columns are the unit vectors

$$e_1 = \begin{bmatrix} 1 \\ 0 \\ \vdots \\ 0 \end{bmatrix}, e_2 = \begin{bmatrix} 0 \\ 1 \\ \vdots \\ 0 \end{bmatrix}, \cdots, e_n = \begin{bmatrix} 0 \\ 0 \\ \vdots \\ 1 \end{bmatrix}. \qquad (5.105)$$

Therefore, for a rank n matrix A (5.105) may be written as n sets of n linear equations

$$Ax_i = e_i, i = 1, \cdots, n \qquad (5.106)$$

where the solution vectors, x_i, are successive columns of the inverse matrix X.

EXAMPLE 5.12

5.7.4 The LU Decomposition

Consider again the Gaussian elimination method with a view to improving its efficiency in applications. Replacing the matrix of coefficients (for the square case) in

(5.65) in rows $i = 2, \cdots, n$, by repeated application of their equivalent expressions from (5.81,5.82) as

$$a_{ij}^{<k>} = m_{ik} a_{kj}^{<k>} + a_{ij}^{<k+1>}, \quad i, j = k+1, \cdots, n. \tag{5.107}$$

The method is demonstrated in the rank 3 system of (5.50) for which (5.69,5.70) apply. Thus the matrix of coefficients in this case is

$$A = \begin{bmatrix} a_{11} & a_{12} & a_{13} \\ m_{21}a_{11} & m_{21}a_{12} + a_{22}^{<2>} & m_{21}a_{13} + a_{23}^{<2>} \\ m_{31}a_{11} & m_{31}a_{12} + m_{32}a_{22}^{<2>} & m_{31}a_{13} + m_{32}a_{23}^{<2>} + a_{33}^{<3>} \end{bmatrix}, \tag{5.108}$$

where, for the last element, (5.73) was applied.

The expression in (5.108) is the product of two matrices

$$A = \begin{bmatrix} 1 & 0 & 0 \\ m_{21} & 1 & 0 \\ m_{31} & m_{32} & 1 \end{bmatrix} \begin{bmatrix} a_{11} & a_{12} & a_{13} \\ 0 & a_{22}^{<2>} & a_{23}^{<2>} \\ 0 & 0 & a_{33}^{<3>} \end{bmatrix}, \tag{5.109}$$

and this is written symbolically as

$$A = LU, \tag{5.110}$$

where L and U are, respectively, lower and upper triangular matrices. To simplify the representation of matrix elements and as a mnemonic for the character of the matrices, we introduce a new notation

$$\ell_{ij} = m_{ij}, \quad i = 2, \cdots, n, \quad j = 1, \cdots, i-1, \tag{5.111}$$

$$u_{ij} = a_{ij}^{<i>}, \quad i = 1, \cdots, n, \quad j = i, \cdots, n, \tag{5.112}$$

where $a_{1j}^{<1>} = a_{1j}$. With this change in notation the matrices for the general rank n case are

$$L = \begin{bmatrix} 1 & 0 & 0 & \cdots & & 0 \\ \ell_{21} & 1 & 0 & \cdots & & \vdots \\ \ell_{31} & \ell_{32} & 1 & \cdots & & \vdots \\ \vdots & & & \ddots & & \vdots \\ \vdots & & & & 1 & 0 \\ \ell_{n1} & \cdots & & \cdots & \ell_{nn-1} & 1 \end{bmatrix}, \tag{5.113}$$

Table 5.5 Doolittle's method for rank 3

Step	row of U	column of L
1	$u_{11} = a_{11}$	$\ell_{21} = \frac{a_{21}}{u_{11}}$
	$u_{12} = a_{12}$	$\ell_{31} = \frac{a_{31}}{u_{11}}$
	$u_{13} = a_{13}$	
2	$u_{22} = a_{22} - \ell_{21}u_{12}$	$\ell_{32} = \frac{a_{32} - \ell_{31}u_{12}}{u_{22}}$
	$u_{23} = a_{23} - \ell_{21}u_{13}$	
3	$u_{33} = a_{33} - \ell_{31}u_{13} - \ell_{32}u_{23}$	

Table 5.6 Doolittle's method for rank n

Step	row of U	column of L
1	$u_{1j} = a_{1j},\, 1 \leq j \leq n$	$\ell_{i1} = \frac{a_{i1}}{u_{11}},\, 2 \leq i \leq n$
k	$u_{kj} = a_{kj} - \sum_{i=1}^{k-1} \ell_{ki}u_{ij},\, k \leq j \leq n$	$\ell_{ik} = \frac{a_{ik} - \sum_{j=1}^{k-1} \ell_{ij}u_{jk}}{u_{kk}}$
n	$u_{nn} = a_{nn} - \sum_{i=1}^{n-1} \ell_{ni}u_{in}$	

and

$$U = \begin{bmatrix} u_{11} & u_{12} & u_{13} & \cdots & \cdots & u_{1n} \\ 0 & u_{22} & u_{23} & \cdots & \cdots & \vdots \\ \vdots & & u_{33} & \cdots & \cdots & \vdots \\ \vdots & & & \ddots & & \vdots \\ \vdots & & & & u_{n-1n-1} & u_{n-1n} \\ 0 & \cdots & \cdots & \cdots & 0 & u_{nn} \end{bmatrix}. \tag{5.114}$$

Doolittle's method provides a simple procedure to construct these matrices by interleaving rows of U and columns of L alternatively. Again, this is first demonstrated for the rank 3 system in Table 5.5 and is then generalized in Table 5.6.

If it happens that $u_{kk} = 0$ at any step, then the matrix A is singular because the determinant $|A|$, as computed from $|LU| = |L||U|$ is zero, because $|U|$ is zero from the discussion following (5.45). Substitution of the LU decomposition for A into the linear system gives

$$LUx = y, \tag{5.115}$$

and this consists of two separate systems of linear equations

$$Lz = y, \tag{5.116}$$

$$Ux = z. \tag{5.117}$$

The importance of the LU decomposition is that, once these matrices have been formed, then the triangular character of the matrices is exploited in solving for x. First

5.7 Matrix Solution Methods

Table 5.7 Crout factorization rank n

Step	column of L	row of U
1	$\ell_{i1} = a_{i1}, 1 \leq i \leq n$	$u_{1j} = \frac{a_{1j}}{u_{11}}, 2 \leq j \leq n$
k	$\ell_{jk} = a_{jk} - \sum_{i=1}^{k-1} \ell_{ji} u_{ik}, k \leq j \leq n$	$u_{ki} = \frac{a_{ki} - \sum_{j=1}^{k-1} \ell_{kj} u_{ji}}{\ell_{kk}}, k < i \leq n$
n	$\ell_{nn} = a_{nn} - \sum_{i=1}^{n-1} \ell_{ni} u_{in}$	

(5.116) is solved for z by forward substitution, similarly to (5.84), (5.85), excepting that the matrix has unity along the diagonal. Therefore, forward substitution in (5.116) gives

$$z_1 = y_1, \tag{5.118}$$

$$z_i = y_i - \sum_{j=1}^{i-1} \ell_{ij} z_j, \, i = 2, \cdots, n, \tag{5.119}$$

and then x is solved for from (5.117) using back substitution

$$x_n = \frac{z_n}{u_{nn}}, \tag{5.120}$$

$$x_i = \frac{z_i - \sum_{j=i+1}^{n} u_{ij} x_j}{u_{ii}}, \, i = n-1, \cdots, 1. \tag{5.121}$$

EXAMPLES 5.13 and 5.14

5.7.5 Crout and Cholesky Decomposition

The LU decomposition of the matrix A in (5.113, 5.114) is not unique and therefore other decompositions of A are possible. Whereas Doolittle's method requires that the diagonal matrix elements of L be unity, in Crout factorization it is the diagonal entries of U that are unity. Therefore, the Crout decomposition is obtained as a suitable modification of Table 5.6 and (5.118)–(5.121) where the divisors are rearranged for the general rank n case as in Table 5.7.

Then forward substitution in (5.116)

$$z_1 = \frac{y_1}{\ell_{11}} \tag{5.122}$$

$$z_i = \frac{y_i - \sum_{j=1}^{i-1} \ell_{ij} z_j}{\ell_{ii}}, \, i = 2, \cdots, n, \tag{5.123}$$

Table 5.8 Cholesky factorization rank n

Step	Diagonal	Off-diagonal
1	$\ell_{11} = \sqrt{a_{11}}$	$\ell_{i1} = \frac{a_{i1}}{\ell_{11}}, 2 \leq i \leq n$
k	$\ell_{kk} = \left[a_{kk} - \sum_{j=1}^{k-1} \ell_{kj}^2\right]^{\frac{1}{2}}$	$\ell_{ik} = \frac{a_{ik} - \sum_{j=1}^{k-1} \ell_{ij}\ell_{kj}}{\ell_{kk}}, k < i \leq n$
n	$\ell_{nn} = \left[a_{nn} - \sum_{i=1}^{n-1} \ell_{ni}^2\right]^{\frac{1}{2}}$	

and solving for x in (5.117) by back substitution

$$x_n = z_n \tag{5.124}$$

$$x_i = z_i - \sum_{j=i+1}^{n} u_{ij} x_j, \; i = n-1, \cdots, 1. \tag{5.125}$$

If A is symmetric and positive definite, then it may be factored into a product of three matrices

$$A = L\Lambda L^T, \tag{5.126}$$

where L is lower triangular with unit matrix elements on the main diagonal and Λ is a diagonal matrix with only positive matrix elements. As a result the property in (5.126) may be manipulated into the form

$$A = (L\sqrt{\Lambda})(\sqrt{\Lambda}L^T)$$
$$= (L\sqrt{\Lambda})(L\sqrt{\Lambda})^T \tag{5.127}$$

where $\sqrt{\Lambda}$ is the diagonal matrix whose matrix elements are the square root of the corresponding matrix elements of Λ. The last result in (5.127) is the Cholesky factorization of A and the lower triangular matrix

$$\mathcal{L} = L\sqrt{\Lambda}, \tag{5.128}$$

no longer has unity along the diagonal. The algebraic scheme for computing the matrix elements of $\mathcal{L}$ is shown in Table 5.8.

Then forward substitution for (5.116) proceeds as before, while back substitution is similar to that for (5.117) with $\mathcal{L}^T$ in place of U and application of (5.120,5.121) using matrix elements from Table 5.8 with $\ell_{ij}^T = \ell_{ji}$ in place of u_{ij}.

EXAMPLES 5.15 and 5.16

5.7.6 Tridiagonal Matrices

Of special interest are matrices where many of the non-zero matrix elements are near to the diagonal. Even if matrix elements further from the diagonal ones are not zero, the matrix may still be diagonally dominant if

5.7 Matrix Solution Methods

$$|a_{ii}| > \sum_{j=1}^{n} |a_{ij}|, j \neq i, 1 \leq i \leq n. \tag{5.129}$$

A matrix has band structure if the matrix elements a_{ij} are zero once the band of index values, $|i - j|$, exceeds some value $k < n$. For example, if $k = 1$, then the matrix has one band of possible non-zero entries above and below the diagonal ones, as in (5.16). Such a system is easily solved in $Ax = y$ by Gaussian elimination and no pivoting is required if the condition in (5.129) holds. In this case the LU decomposition takes a particularly simple form

$$LU = \begin{bmatrix} 1 & 0 & 0 & \cdots & & \cdots & 0 \\ \ell_{21} & 1 & 0 & \cdots & & \cdots & \vdots \\ 0 & \ell_{32} & 1 & \cdots & & \cdots & \vdots \\ \vdots & & \ddots & & & & \vdots \\ \vdots & & & 1 & 0 & & \\ 0 & \cdots & \cdots & \cdots & \ell_{nn-1} & 1 \end{bmatrix} \begin{bmatrix} u_{11} & u_{12} & 0 & \cdots & & \cdots & 0 \\ 0 & u_{22} & u_{23} & \cdots & & \cdots & \vdots \\ \vdots & & u_{33} & \cdots & & \cdots & \vdots \\ \vdots & & & \ddots & & & \vdots \\ \vdots & & & & & u_{n-1n-1} & u_{n-1n} \\ 0 & \cdots & \cdots & \cdots & & 0 & u_{nn} \end{bmatrix}, \tag{5.130}$$

or

$$LU = \begin{bmatrix} u_{11} & u_{12} & 0 & \cdots & & \cdots & 0 \\ \ell_{21}u_{11} & \ell_{21}u_{12} + u_{22} & u_{23} & \cdots & & \cdots & \vdots \\ \vdots & \ell_{32}u_{22} & \ell_{32}u_{23} + u_{33} & \cdots & & & \vdots \\ \vdots & & & \ddots & & & \vdots \\ \vdots & & & & & & u_{n-1n} \\ 0 & \cdots & & \cdots & & \ell_{nn-1}u_{n-1n-1} & \ell_{nn-1}u_{n-1n} + u_{nn} \end{bmatrix}. \tag{5.131}$$

Then the unknowns ℓ_{ij}, u_{ij} in (5.131) are solved for by equating to the respective matrix elements of (5.16) as follows:

$$u_{11} = a_{11} \tag{5.132}$$
$$u_{ii+1} = a_{ii+1}, i = 1, \cdots, n-1, \tag{5.133}$$

with

$$\ell_{ii-1} = \frac{a_{ii-1}}{u_{i-1i-1}}, \tag{5.134}$$

$$u_{ii} = a_{ii} - \ell_{ii-1}u_{i-1i}, \tag{5.135}$$

applied alternately for $i = 2, \cdots, n-1$ and finally (5.135) is used for $i = n$. The vector z is obtained by forward substitution in $Lz = y$ as a special case of (5.118, 5.119)

$$z_1 = y_1, \tag{5.136}$$
$$z_i = y_i - \ell_{ii-1}z_{i-1}, i = 2, \cdots, n, \tag{5.137}$$

and then x is solved for by back substitution as a special case of (5.120), (5.121)

$$x_n = \frac{z_n}{u_{nn}} \tag{5.138}$$

$$x_i = \frac{z_i - u_{ii+1}x_{i+1}}{u_{ii}}, i = n-1, \cdots, 1. \tag{5.139}$$

Because of the special structure of the tridiagonal matrix system, a simplified notation is often introduced. For example, define the three diagonals of matrix elements in A as

$$\alpha_i = a_{ii-1}, i = 2, \cdots, n, \tag{5.140}$$
$$\beta_i = a_{ii}, i = 1, \cdots, n, \tag{5.141}$$
$$\gamma_i = a_{ii+1}, i = 1, \cdots, n-1, \tag{5.142}$$

then the matrix L has matrix elements below the main diagonal defined by

$$\lambda_i = \frac{\alpha_i}{\beta_{i-1}}, i = 2, \cdots, n, \tag{5.143}$$

while those along the main diagonal of matrix U are

$$\mu_1 = \beta_1 \tag{5.144}$$
$$\mu_i = \beta_i - \lambda_i \gamma_{i-1}, i = 2, \cdots, n, \tag{5.145}$$

and the matrix elements above the main diagonal in U are γ_i as defined in (5.142). This should be obvious from application of STEP k of Table 5.6 to the tridiagonal case. Note that (5.143) to (5.145) have to be applied alternately, as in the case of (5.134), (5.135).

EXAMPLES 5.17 and 5.18

5.7.7 Gauss-Jordan Elimination

In the Gaussian elimination method, at step k the unknown x_k is eliminated from all subsequent equations. This method may be modified by also eliminating the unknown from the preceding equations when the kth equation is divided throughout by the coefficient of x_k. This is known as the Gauss-Jordan method.

The method is demonstrated in the rank three system of (5.50) by back elimination in (5.68) after the first forward elimination with the new multiplier

$$w_{12} = \frac{a_{12}}{a_{22}^{<2>}}. \tag{5.146}$$

5.7 Matrix Solution Methods

Subtraction of the second row of (5.68), after multiplying it by w_{12}, from the first row, eliminates the coefficient of x_2 and updates y_1 and a_{13} as follows:

$$a_{13}^{<2>} = a_{13} - w_{12} a_{23}^{<2>}, \quad (5.147)$$

$$y_1^{<2>} = y_1 - w_{12} y_2^{<2>}. \quad (5.148)$$

Then, in place of (5.68), the modified equations are

$$a_{11} x_1 + 0 + a_{13}^{<2>} x_3 = y_1^{<2>}$$
$$a_{22}^{<2>} x_2 + a_{23}^{<2>} x_3 = y_2^{<2>}$$
$$a_{32}^{<2>} x_2 + a_{33}^{<2>} x_3 = y_3^{<2>}. \quad (5.149)$$

Then, as in (5.71), where a multiplier m_{32} is used in forward elimination to give (5.72), here new multipliers are formed from

$$w_{i3} = \frac{a_{i3}^{<2>}}{a_{33}^{<3>}}, \quad i = 1, 2, \quad (5.150)$$

from which $a_{13}^{<2>}, a_{23}^{<2>}$, are eliminated to give a second set of modified equations

$$a_{11} x_1 + 0 + 0 = y_1^{<3>}$$
$$a_{22}^{<2>} x_2 + 0 = y_2^{<3>}$$
$$a_{33}^{<3>} x_3 = y_3^{<3>}, \quad (5.151)$$

where

$$y_i^{<3>} = y_i^{<2>} - w_{i3} y_3^{<3>}, \quad i = 1, 2. \quad (5.152)$$

The result of (5.151), when represented in matrix form, is

$$A^{<3>} x = y^{<3>}, \quad (5.153)$$

and in this case the matrix $A^{<3>}$ is diagonal, with the solution, x, now easily obtained as

$$x_i = \frac{y_i^{<3>}}{a_{ii}^{<3>}}, \quad i = 1, 2, 3. \quad (5.154)$$

The Gauss-Jordan elimination is a two-step process where the first step is forward elimination of unknowns and the second is back eliminations of the same unknowns. If the system is of rank n, then this method approximately doubles the arithmetic work load of Gaussian elimination that grows as $\mathcal{O}(n^3)$, but it does remove the need for back substitution.

In the general case of the Gauss-Jordan elimination algorithm, on defining $y_i^{<1>} = y_i$ and $a_{ij}^{<1>} = a_{ij}$, at step k unknowns x_1 to x_{k-1} are eliminated and the linear system is reduced to the form

$$a_{11}^{<1>}x_1 + 0 + \cdots + a_{1k}^{<k-1>}x_k + \cdots + a_{1n}^{<k-1>}x_n = y_1^{<k-1>}$$
$$a_{22}^{<2>}x_2 + \cdots + a_{2k}^{<k-1>}x_k + \cdots + a_{2n}^{<k-1>}x_n = y_2^{<k-1>}$$
$$\vdots$$
$$a_{kk}^{<k>}x_k + \cdots + a_{kn}^{<k>}x_n = y_k^{<k>}$$
$$\vdots$$
$$a_{nk}^{<k>}x_k + \cdots + a_{nn}^{<k>}x_n = y_n^{<k>}. \tag{5.155}$$

Because of back elimination, the non-zero coefficients above the kth equation, with the exception of the diagonal ones, have been modified $k - 2$ times in previous steps. At the kth step, backward elimination is performed by forming multipliers

$$w_{ik} = \frac{a_{ik}^{<k-1>}}{a_{kk}^{<k>}}, i = 1, \cdots, k-1, \tag{5.156}$$

and the coefficients of x_k are eliminated for equations 1 to $k - 1$. At the same time coefficients for x_{k+1} to x_n are updated, together with the right-hand side of (5.155), by application of

$$a_{ij}^{<k>} = a_{ij}^{<k-1>} - w_{ik}a_{kj}^{<k>}, j = k+1, \cdots, n, \tag{5.157}$$
$$y_i^{<k>} = y_i^{<k-1>} - w_{ik}y_k^{<k>}, \tag{5.158}$$

for $i = 1, \cdots, k - 1$. Thus the Gauss-Jordan method modifies the matrix A through a sequence of transformations

$$A^{<1>} \to A^{<2>} \to \cdots \to A^{<n-1>} \to A^{<n>}, \tag{5.159}$$

where the limit of the sequence $A^{<n>}$, is a diagonal matrix. Once the transformations have been completed, then the solution in the rank n case is

$$x_i = \frac{y_i^{<n>}}{a_{ii}^{<n>}}, i = 1, \cdots, n. \tag{5.160}$$

One of the more important applications of the Gauss-Jordan elimination method is in the computation of the inverse of a matrix A. If this method is used for the matrix obtained by augmenting A with the identity matrix

$$\begin{bmatrix} a_{11} & a_{12} & \cdots & a_{1n} & | & 1 & 0 & \cdots & 0 \\ a_{21} & a_{22} & \cdots & a_{2n} & | & 0 & 1 & \cdots & 0 \\ \vdots & \vdots & \ddots & \vdots & | & \vdots & \vdots & \ddots & \vdots \\ a_{n1} & a_{n2} & \cdots & a_{nn} & | & 0 & 0 & \cdots & 1 \end{bmatrix} \tag{5.161}$$

then application of elimination steps to all rows of the augmented matrix results in a left-hand half that is diagonal. After dividing each row by the corresponding entry

$a_{kk}^{<k>}, k = 1, \cdots, n$, the right-hand part of the augmented matrix is the inverse matrix A^{-1}.

5.8 Matrix Norms, Condition Number and Stability

The matrix methods discussed thus far are all algebraically exact in the sense that given infinite precision arithmetic, they would produce exact results because the matrix A and its transformations are computed exactly at each step. However, because of the finite word length of computing machines (as discussed in Chap. 1), round-off error will accumulate. This is one reason why a measure of error in the matrix transformations is required from step to step. There is also another reason for measuring such error. Direct methods, such as Gaussian, or Gauss-Jordan, elimination represent one class of method for solving linear systems. Another important class is iterative techniques that are discussed in the following sections. Iterative methods replace the exact matrix A by an algebraic approximation and proceed to improve this approximation from one step to the next. The progress of the calculation in an iterative method therefore requires a measure of the size of the change between iterations. The appropriate measure of such differences is the norm of a matrix that is defined in this section after a brief review of some results from Chap. 4, as they apply to vectors and matrices.

A vector x, of rank n, is a member of a set that spans the finite-dimensional space $\mathbf{K}^n$, consisting of complex elements of finite discrete sets $x^T = [x_1, x_2, \cdots, x_n]$ and therefore, possible norms are those that are listed in Section 4.4, as follows:

ℓ^2, or Euclidean norm

$$\| x \|_2 = \left[\sum_{i=1}^{n} | x_i |^2 \right]^{\frac{1}{2}}, \qquad (5.162)$$

uniform norm

$$\| x \|_\infty = \max_{i=1,\cdots,n} | x_i |, \qquad (5.163)$$

ℓ^1 norm

$$\| x \|_1 = \sum_{i=1}^{n} | x_i |, \qquad (5.164)$$

ℓ^p norm

$$\| x \|_p = \left[\sum_{i=1}^{n} | x_i |^p \right]^{\frac{1}{p}}. \qquad (5.165)$$

Because the uniform norm is a consequence of the limit $p \to \infty$, it is therefore sometimes denoted by ℓ^∞. Each choice of a norm requires boundedness (usually satisfied in the rank n case) and this is defined by

$$\sum_{i=1}^{n} |x_i|^p < \infty. \tag{5.166}$$

Any one of these norms could be used to measure the length of a vector in $\mathbf{K}^n$ but convergence properties are different for each choice. Suppose that the vector $x = x^{<0>} - x^{<m>}$ measures the difference between an element of a sequence of vectors $\{x^{<m>}\}_1^\infty$, and the limit vector $x^{<0>}$. Then each norm gives a different value for the distance, or "size" of x. Thus, whereas the ℓ^2 norm would give the distance

$$\| x^{<0>} - x^{<m>} \|_2 = \left[\sum_{i=1}^{n} | x_i^{<0>} - x_i^{<m>} |^2 \right]^{\frac{1}{2}}, \tag{5.167}$$

the uniform norm selects only the maximum of the difference of any two components

$$\| x^{<0>} - x^{<m>} \|_\infty = \max_{i=1,\cdots,n} | x_i^{<0>} - x_i^{<m>} |. \tag{5.168}$$

Each of these norms satisfies the axioms and properties discussed in Section 4.3 and convergence in each norm is defined in terms of (4.15,4.16). Of particular significance is convergence in the uniform norm, that, on requiring (5.168), converges to zero. This also ensures that

$$\lim_{m \to \infty} x_i^{<m>} \to x_i^{<0>}, \; i = 1, \cdots, n, \tag{5.169}$$

which is known as "coordinate-wise" convergence.

There exists a simple relationship between the ℓ^∞ and the ℓ^2 norms for a rank n vector

$$\| x \|_\infty \leq \| x \|_2 \leq \sqrt{n} \, \| x \|_\infty, \tag{5.170}$$

that follows, if it is assumed that for some $1 \leq j \leq n$,

$$\begin{aligned} \| x \|_\infty^2 &= | x_j |^2 \\ &= x_j^2 \\ &\leq \sum_{i=1}^{n} x_i^2 \\ &\leq n x_j^2 \\ &= n \| x \|_\infty^2, \end{aligned} \tag{5.171}$$

and then (5.170) follows on taking the square root and applying (5.162).

The distance between two rank n matrices A, B may be determined in terms of a matrix norm $||A - B||$. The norm of a matrix, $||A||$, satisfies axioms and properties similar to those of Sect. 4.3, with the modification that AXIOM 4.3.I, ||A||=0, implies that the matrix has only zero entries. An additional property of the matrix norm is

$$\| AB \| \leq \| A \| \| B \|, \tag{5.172}$$

5.8 Matrix Norms, Condition Number and Stability

and this may be applied to define the matrix norm for the special case of $B = x$, a vector with norm $||x|| = 1$. From the properties of the matrix-vector product in Table 5.3, Ax is seen to be a vector. Therefore any vector norm $||.||$ in (5.162) to (5.165) may be applied to define the matrix norm. From this argument, (5.172) becomes

$$|| Ax || \leq || A |||| x ||, \qquad (5.173)$$

and, if it is assumed that $||x|| = 1$ (otherwise first normalize with $x/||x||$), then this implies

$$|| Ax || \leq C, \qquad (5.174)$$

where C is a constant. The question then is: what is the smallest value, C_{min}, of C that still satisfies the inequality of (5.174)? This C_{min}, would be the maximum value of the left-hand side, $||Ax||$, and therefore the norm of the matrix is written as $||A|| = C_{min}$ and is defined as

$$|| A || = \max_{||x||=1} || Ax ||, \qquad (5.175)$$

and the matrix norm is bounded if $||A||$ is finite. The importance of the matrix norm is in the study of convergence of sequences of vectors that converge in norm

$$|| x^{<0>} - x^{<m>} || \to 0, \qquad (5.176)$$

$$\lim_{m \to \infty} x^{<m>} \to x^{<0>}, \qquad (5.177)$$

for which application of (5.173), with $x = x^{<0>} - x^{<m>}$, gives

$$|| Ax^{<0>} - Ax^{<m>} || \leq || A |||| x^{<0>} - x^{<m>} ||. \qquad (5.178)$$

In view of convergence in norm of the sequence of vectors in (5.177), the left-hand side of (5.178) must also converge if $||A||$ is bounded. However, the rate of convergence is determined by the norm $||.||$ that is applied and also the magnitude of $||A||$.

To show how (5.178) appears in specific norms, consider three of the vector norms for systems defined by $Ax^{<0>}$ and $Ax^{<m>}$ with A having matrix elements a_{ij}. The first such norm is the Euclidean norm of (5.162)

$$|| Ax^{<0>} - Ax^{<m>} ||_2 = \left[\sum_{i=1}^{n} \left\{ \sum_{j=1}^{n} a_{ij} \left(x_j^{<0>} - x_j^{<m>} \right) \right\}^2 \right]^{\frac{1}{2}},$$

$$\leq \left[\sum_{i=1}^{n} \left\{ \sum_{j=1}^{n} |a_{ij}|^2 \times \sum_{j=1}^{n} \left(x_j^{<0>} - x_j^{<m>} \right)^2 \right\} \right]^{\frac{1}{2}},$$

$$\leq \left[\sum_{i=1}^{n}\sum_{j=1}^{n}|a_{ij}|^2\right]^{\frac{1}{2}} \| x^{<0>} - x^{<m>} \|_2, \qquad (5.179)$$

and the Euclidean matrix norm of A, sometimes called the Frobenius norm, is

$$\| A \|_e = \left[\sum_{i=1}^{n}\sum_{j=1}^{n}|a_{ij}|^2\right]^{\frac{1}{2}}, \qquad (5.180)$$

which follows from the comparison of (5.179) and (5.173). Whereas for a vector, the ℓ^2 and Euclidean norms are the same and follow from the vector product $x^T x$, for a matrix they are different. The ℓ^2 norm of a matrix is related to $A^T A$ which does not equal the result of (5.180) and requires a development of the eigenvalue problem that is not discussed in depth in this book, although a case study is shown in Chap. 9.

Consider the case of the uniform norm of (5.163) for which

$$\| Ax^{<0>} - Ax^{<m>} \|_\infty = \max_{i=1,\cdots,n} |\sum_{j=1}^{n} a_{ij}\left(x_j^{<0>} - x_j^{<m>}\right)|,$$

$$\leq \max_{i=1,\cdots,n} \sum_{j=1}^{n} |a_{ij}||x_j^{<0>} - x_j^{<m>}|,$$

$$\leq \max_{i=1,\cdots,n} \sum_{j=1}^{n} |a_{ij}| \times \max_{i=1,\cdots,n} |x_j^{<0>} - x_j^{<m>}|,$$

$$\leq \max_{i=1,\cdots,n} \sum_{j=1}^{n} |a_{ij}| \times \| x^{<0>} - x^{<m>} \|_\infty, \qquad (5.181)$$

and the uniform matrix norm of A is

$$\| A \|_\infty = \max_{i=1,\cdots,n} \sum_{j=1}^{n} |a_{ij}|. \qquad (5.182)$$

The final example is the ℓ^1 norm of (5.164) that gives

$$\| Ax^{<0>} - Ax^{<m>} \|_1 = \sum_{i=1}^{n} |\sum_{j=1}^{n} a_{ij}\left(x_j^{<0>} - x_j^{<m>}\right)|,$$

$$\leq \sum_{i=1}^{n}\sum_{j=1}^{n} |a_{ij}||x_j^{<0>} - x_j^{<m>}|,$$

5.8 Matrix Norms, Condition Number and Stability

$$\leq \left(\max_{j=1,\cdots,n} \sum_{i=1}^{n} |a_{ij}| \right) \sum_{j=1}^{n} |x_j^{<0>} - x_j^{<m>}|,$$

$$\leq \left(\max_{j=1,\cdots,n} \sum_{i=1}^{n} |a_{ij}| \right) \| x^{<0>} - x^{<m>} \|_1, \quad (5.183)$$

and the ℓ^1 norm of matrix A is

$$\| A \|_1 = \max_{j=1,\cdots,n} \sum_{i=1}^{n} |a_{ij}|. \quad (5.184)$$

An important application of the matrix norm is in determining the stability of the unperturbed linear system

$$Ax^{<u>} = y^{<u>}, \quad (5.185)$$

with respect to a small perturbation of the right-hand side, such that the perturbed system is

$$Ax^{<p>} = y^{<p>}. \quad (5.186)$$

In other words, how is the change in the solution $x^{<u>} - x^{<p>}$ related to the change $y^{<u>} - y^{<p>}$ in the right-hand side of the matrix system? To investigate this, subtract (5.186) from (5.185), then multiply from the left by A^{-1}, the inverse of A, to obtain

$$x^{<u>} - x^{<p>} = A^{-1}(y^{<u>} - y^{<p>}). \quad (5.187)$$

On applying the norm to both sides

$$\|x^{<u>} - x^{<p>}\| \leq \|A^{-1}\| \|y^{<u>} - y^{<p>}\|, \quad (5.188)$$

and multiplication by $\|A\| / \{\|x^{<u>}\| \|A\|\}$ gives

$$\frac{\|x^{<u>} - x^{<p>}\|}{\|x^{<u>}\|} \leq \frac{\|A\| \|A^{-1}\| \|y^{<u>} - y^{<p>}\|}{\|x^{<u>}\| \|A\|}. \quad (5.189)$$

In the denominator of the right-hand side, the result $\|y^{<u>}\| \leq \|A\| \|x^{<u>}\|$ from (5.185), is applied and it follows that for non-zero $\|x^{<u>}\|$, $\|y^{<u>}\|$ and A non-singular

$$\frac{\|x^{<u>} - x^{<p>}\|}{\|x^{<u>}\|} \leq \|A\| \|A^{-1}\| \frac{\|y^{<u>} - y^{<p>}\|}{\|y^{<u>}\|}. \quad (5.190)$$

The ratio on the left-hand side is the relative change in the solution that is clearly bounded by the product of the relative change in the right-hand side of (5.185,5.186). The number

$$C(A) = \|A\| \|A^{-1}\|, \quad (5.191)$$

is called the condition number of the matrix A. If $C(A)$ is large, then small perturbations in the right-hand side of a system of linear equations may lead to large changes in the corresponding solution. The source of the perturbations may be round-off error or an approximation scheme such as that due to an iterative method of the type discussed in the next section. Linear systems of equations for which $C(A)$ is large are said to be ill-conditioned and the solution method can become unstable. The converse is the case when the condition number is small and the linear system is well conditioned.

EXAMPLES 5.19–5.21

5.9 Iterative Solution Methods

When the rank n of the linear system of equations to be solved becomes large, the cost of applying direct solution methods can become prohibitive because the number of operations scales as $\mathcal{O}(n^3)$. In these circumstances, alternative solution methods are sought and iterative algorithms present attractive options because the arithmetic operations are simpler and fewer in number than in direct methods.

The idea behind all iterative methods is to approximate the matrix A and solve the system of equations $Ax = y$ by generating a sequence of approximants $\{x^{<m>}\}_1^\infty$ that converge in norm to the solution. An iteration algorithm applies a suitable factorization of the matrix A into a sum of matrices of special structure, by analogy with the product factorizations of Sects. 5.7.4 and 5.7.5. As an example, consider the form obtained when matrix A is factored into a sum of three separate matrices

$$(L + D + U)x = y, \tag{5.192}$$

where L, U are lower and upper triangular matrices

$$L = \begin{bmatrix} 0 & 0 & 0 & \cdots & & \cdots & 0 \\ a_{21} & 0 & 0 & \cdots & & \cdots & 0 \\ a_{31} & a_{32} & 0 & \cdots & & \cdots & \vdots \\ \vdots & & & \ddots & & & \vdots \\ \vdots & & & & & 0 & 0 \\ a_{n1} & \cdots & & \cdots & \cdots & a_{nn-1} & 0 \end{bmatrix}, \tag{5.193}$$

and

$$U = \begin{bmatrix} 0 & a_{12} & a_{13} & \cdots & \cdots & a_{1n} \\ 0 & 0 & a_{23} & \cdots & \cdots & \vdots \\ \vdots & & 0 & \cdots & \cdots & \vdots \\ \vdots & & & \ddots & & \vdots \\ \vdots & & & & 0 & a_{n-1n} \\ 0 & \cdots & \cdots & \cdots & 0 & 0 \end{bmatrix}, \tag{5.194}$$

5.9 Iterative Solution Methods

while the matrix D has only the diagonal entries of A as non-zero entries

$$D = \begin{bmatrix} a_{11} & 0 & 0 & \cdots & & 0 \\ 0 & a_{22} & 0 & \cdots & & \vdots \\ \vdots & & a_{33} & \cdots & & \vdots \\ \vdots & & & \ddots & & \vdots \\ & & & & a_{n-1n-1} & 0 \\ 0 & \cdots & \cdots & \cdots & 0 & a_{nn} \end{bmatrix}. \tag{5.195}$$

Then, if the triangular matrices of (5.192) are moved to the right-hand side, the system has the matrix form

$$Dx = -(L+U)x + y, \tag{5.196}$$

and on left multiplication by D^{-1} this becomes

$$x = -D^{-1}(L+U)x + D^{-1}y, \tag{5.197}$$

where the vector

$$d = D^{-1}y, \tag{5.198}$$

is a constant matrix and

$$D' = -D^{-1}(L+U), \tag{5.199}$$

has no diagonal entries. Note that D^{-1} is diagonal with matrix elements that are the inverse of the corresponding entries in D. To define an iterative solution, suppose that on the right-hand side of (5.197) we substitute $x = 0$ to obtain the first estimate for x as $x^{<1>} = d$. Then (5.197), with the notation of (5.198), (5.199) may be applied as an iteration to generate the sequence of vectors

$$x^{<m>} = D'x^{<m-1>} + d, \tag{5.200}$$

for $m = 2, 3, \cdots$. The method of (5.200) is known as the Jacobi iteration and requires that the diagonal entries of the original matrix A are all non-zero. The Jacobi iteration updates all components of the vector $x^{<m-1>}$ simultaneously, as is shown in a simple rank 3 example

$$\begin{aligned} x_1^{<m>} &= \frac{y_1 - a_{12}x_2^{<m-1>} - a_{13}x_3^{<m-1>}}{a_{11}}, \\ x_2^{<m>} &= \frac{y_2 - a_{21}x_1^{<m-1>} - a_{23}x_3^{<m-1>}}{a_{22}}, \\ x_3^{<m>} &= \frac{y_3 - a_{31}x_1^{<m-1>} - a_{32}x_2^{<m-1>}}{a_{33}}. \end{aligned} \tag{5.201}$$

However, the updated value of $x_1^{<m>}$, available from (5.201), may be substituted into the expression for $x_2^{<m>}$, for a better estimate than the previous value. Similarly, an improved updated value of $x_3^{<m>}$ may be solved for by substituting $x_1^{<m>}$, $x_2^{<m>}$ on the right-hand side. This modified iteration uses the system of equations

$$x_1^{<m>} = \frac{y_1 - a_{12}x_2^{<m-1>} - a_{13}x_3^{<m-1>}}{a_{11}},$$

$$x_2^{<m>} = \frac{y_2 - a_{21}x_1^{<m>} - a_{23}x_3^{<m-1>}}{a_{22}},$$

$$x_3^{<m>} = \frac{y_3 - a_{31}x_1^{<m>} - a_{32}x_2^{<m>}}{a_{33}}, \quad (5.202)$$

and if (5.202) is arranged such that the updated vector $x^{<m>}$ is on the left-hand side, with the vector $x^{<m-1>}$ on the right-hand side, then the matrix form of the rank three system is

$$\begin{bmatrix} a_{11} & 0 & 0 \\ a_{21} & a_{22} & 0 \\ a_{31} & a_{32} & a_{33} \end{bmatrix} \begin{bmatrix} x_1^{<m>} \\ x_2^{<m>} \\ x_3^{<m>} \end{bmatrix} = \begin{bmatrix} y_1 \\ y_2 \\ y_3 \end{bmatrix} - \begin{bmatrix} 0 & a_{12} & a_{13} \\ 0 & 0 & a_{23} \\ 0 & 0 & 0 \end{bmatrix} \begin{bmatrix} x_1^{<m-1>} \\ x_2^{<m-1>} \\ x_3^{<m-1>} \end{bmatrix}. \quad (5.203)$$

This modified form is known as the Gauss-Seidel iteration and has the general form in the rank n case

$$(L + D)x = -Ux + y, \quad (5.204)$$

from which the Gauss-Seidel iteration is obtained as

$$x^{<m>} = Gx^{<m-1>} + g, \quad (5.205)$$

where the constant vector g is defined as

$$g = (L + D)^{-1}y, \quad (5.206)$$

and the matrix G is

$$G = -(L + D)^{-1}U. \quad (5.207)$$

Both Gauss-Jordan elimination and Gauss-Seidel iteration are examples of methods that result from splitting the original matrix A and require the assumption that all diagonal entries are non-zero so that D or $L + D$ are non-singular matrices. Such splittings are not unique and may be chosen any number of ways. Another example would be

$$(S + A - S)x = y, \quad (5.208)$$

that would give the iteration

$$x^{<m>} = S'x^{<m-1>} + s, \quad (5.209)$$

where the constant vector is defined by

$$s = S^{-1}y, \tag{5.210}$$

and the matrix S' is

$$S' = -S^{-1}(A - S), \tag{5.211}$$

with a choice for a non-singular matrix S that makes for easy solution of the inverse.

EXAMPLES 5.22 to 5.25

5.10 Convergence Criteria

Whichever iteration method is chosen, a suitable test must be performed to determine if convergence has been achieved. Because the exact result is not known, the difference between successive iterates may be inspected in an appropriate norm and the iteration terminated if

$$\frac{||x^{<m>} - x^{<m-1>}||}{||x^{<m>}||} \leq \epsilon, \tag{5.212}$$

for a suitable choice of the relative error ϵ. However, (5.212) offers no information on the distance of the mth iterate from the exact result. To establish convergence criteria, let $x^{<0>}$ be the limit of the sequence that equals the exact solution x of $Ax = y$, and define the residual vector for the mth approximant as

$$\begin{aligned} r^{<m>} &= Ax^{<0>} - Ax^{<m>}, \\ &= y - y^{<m>}, \end{aligned} \tag{5.213}$$

where $y^{<m>} = Ax^{<m>}$. Then a bound on the relative distance of $x^{<m>}$ from the exact solution is given by the inequality (5.190), that, on application of (5.213) with the present notation, becomes

$$\frac{||x - x^{<m>}||}{||x||} \leq ||A|| \, ||A^{-1}|| \frac{||r^{<m>}||}{||y||}. \tag{5.214}$$

From this it may be concluded that if the condition number of (5.191) is large, then convergence of an iterative scheme could become problematic and even if convergence exists it may be very slow. Therefore, it is important to determine if the iterative sequence $\{x^{<m>}\}_1^\infty$ does converge to the exact result and, if so, at which rate?

To investigate these convergence questions, consider the general form of the iterative algorithms discussed in Sect. 5.9

$$x^{<m>} = Tx^{<m-1>} + t, \tag{5.215}$$

where $T = D', G$, or S' for (5.200, 5.205, 5.209), respectively and similarly the constant vector $t = d, g$, or s for (5.198, 5.206, 5.210). The sequence generated by (5.215) then satisfies the equations

$$x^{<2>} = Tx^{<1>} + t,$$
$$x^{<3>} = Tx^{<2>} + t,$$
$$x^{<4>} = Tx^{<3>} + t,$$
$$\vdots$$
$$x^{<0>} = Tx^{<0>} + t, \tag{5.216}$$

where $x^{<0>}$, the limit of the sequence, or solution, also satisfies (5.215). Therefore a simple manipulation of the last equation, shows that

$$x^{<0>} = (I - T)^{-1}t, \tag{5.217}$$

where I is the identity matrix. In other words, if the inverse matrix of $(I-T)$ is known, then the solution is also known from (5.217). However, solving for this inverse is as difficult as solving the system of linear equations (5.216). Nevertheless, some properties of $(I - T)^{-1}$ may be determined to establish answers to the convergence questions raised above for the iteration (5.215).

Suppose that $||T|| \leq \tau < 1$ and apply (5.172) to determine that $||T^k|| \leq ||T||^k$. Then consider the matrix V, defined as the convergent power series of the matrix T

$$V = \sum_{k=0}^{\infty} T^k, \tag{5.218}$$

then apply it to compute the product

$$V(I - T) = VI - VT,$$
$$= (I + T + \cdots + T^k + \cdots) - (T + T^2 + \cdots + T^{k+1} + \cdots),$$
$$= I, \tag{5.219}$$

and the same result holds for the product $(I - T)V$. Therefore, the definition (5.28) shows that V is the inverse of $(I - T)$ in (5.219) and consequently

$$(I - T)^{-1} = \sum_{k=0}^{\infty} T^k. \tag{5.220}$$

Now the norm of this inverse is computed using the result $||T^k|| \leq ||T||^k$

$$||(I - T)^{-1}|| = ||\sum_{k=0}^{\infty} T^k||,$$
$$\leq \sum_{k=0}^{\infty} ||T^k||,$$

5.10 Convergence Criteria

$$\leq \sum_{k=0}^{\infty} ||T||^k,$$

$$\leq \sum_{k=0}^{\infty} \tau^k,$$

$$\leq \frac{1}{1-\tau}, \tag{5.221}$$

where the last result is the known sum of the power series in τ. The relationship between the iterative series method of (5.215) that generates powers of T and the inverse $(I - T)^{-1}$ is now apparent in (5.220). To demonstrate this equivalence, we now proceed to show that if the series of (5.220) is convergent, then the iteration of (5.215) converges to the unique solution of (5.216) for any initial approximation $x^{<1>}$. To show this, collect the results of (5.216) into the expression

$$\begin{aligned} x^{<m>} &= t + Tx^{<m-1>}, \\ &= t + T\left(t + Tx^{<m-2>}\right), \\ &= t + Tt + \cdots + T^{m-2}t + T^{m-1}x^{<1>}, \quad m = 2, 3, \cdots \end{aligned} \tag{5.222}$$

and in the limit $m \to \infty$, this gives the exact result of (5.217)

$$\begin{aligned} x^{<0>} &= \lim_{m \to \infty} x^{<m>} \\ &= \sum_{k=0}^{\infty} T^k t, \\ &= (I - T)^{-1} t. \end{aligned} \tag{5.223}$$

Because $x^{<0>}$ is a solution it must also satisfy (5.222)

$$x^{<0>} = t + Tt + \cdots + T^{m-2}t + T^{m-1}x^{<0>}, \quad m = 2, 3, \cdots \tag{5.224}$$

then subtracting (5.222) from (5.224) gives the result

$$x^{<0>} - x^{<m>} = T^{m-1}x^{<0>} - T^{m-1}x^{<1>}, \quad m = 2, 3, \cdots, \tag{5.225}$$

and from this a bound may be established on the error of the $(m-1)$st iterate as

$$||x^{<0>} - x^{<m>}|| \leq ||T^{m-1}||\, ||x^{<0>} - x^{<1>}||, \quad m = 2, 3, \cdots. \tag{5.226}$$

This bound is not very useful in practice because the exact (unknown) solution $x^{<0>}$ appears on the right-hand side. A further simplification is possible by introducing a vector $z = x^{<0>} - x^{<1>}$ and applying (5.216) to obtain the following result

$$(I-T)z = z - Tz,$$
$$= x^{<0>} - Tx^{<0>} - \left(x^{<1>} - Tx^{<1>}\right),$$
$$= t + Tx^{<1>} - x^{<1>},$$
$$= x^{<2>} - x^{<1>}, \qquad (5.227)$$

and therefore $z = (I-T)^{-1}\left(x^{<2>} - x^{<1>}\right)$. When this expression for $z = x^{<0>} - x^{<1>}$ is substituted into the right-hand side of (5.226), the final result is obtained as

$$||x^{<0>} - x^{<m>}|| \leq ||(I-T)^{-1}||\,||T^{m-1}||\,||x^{<2>} - x^{<1>}||, \; m = 2, 3, \cdots, \qquad (5.228)$$

and in view of the bound (5.221) this result may be written as

$$||x^{<0>} - x^{<m>}|| \leq \frac{\tau^{m-1}}{1-\tau}||x^{<2>} - x^{<1>}||, \; m = 2, 3, \cdots. \qquad (5.229)$$

This last result answers the convergence questions concerning the iteration of (5.225). Clearly, convergence is only possible if $||T|| \leq \tau < 1$ and the relative rate of convergence is given by the simple ratio in τ on the right-hand side. The choice of matrix norm is open but could be any one of the possibilities discussed in Sect. 5.8. Observe the remarkable feature of (5.229) that if $\tau < 1$ then, the iteration (5.215) converges to the exact result $x^{<0>}$ for any choice of the initial iterate.

EXAMPLES 5.26–5.29

5.11 Relaxation

For an iterative scheme, Sect. 5.10 developed a quantitative criterion that specifies whether convergence exists and also the rate at which it occurs. When convergence does exist it may be either too slow or too rapid and then, for these respective cases, the iterative method should be either accelerated or damped. This modification to an iterative algorithm is called relaxation.

To describe relaxation, consider the difference between the old and the new estimate for the solution at the mth iteration

$$\Delta x^{<m>} = x^{<m>} - x^{<m-1>}, \qquad (5.230)$$

which is the distance between $x^{<m>}$ and $x^{<m-1>}$ in the n−dimensional space spanned by the rank n vectors. If it is convergent, then the iteration method will produce a correction $\Delta x^{<m>}$ that either overshoots, or does the opposite and undershoots. In the case of overshoot, instead of applying the full correction $\Delta x^{<m>}$, relaxation uses only a smaller fraction of it so that at the mth iterate the approximant is chosen as

$$x^{<m_r>} = x^{<m>} + \mu \Delta x^{<m>}, \qquad (5.231)$$

for a constant $\mu < 1$. The converse is the case of undershoot, or slow convergence to the exact solution when the relaxation applies (5.231) for $\mu > 1$, with the assumption that the vector $\Delta x^{<m>}$ does point to the exact solution in the n-dimensional space. Obviously, the precise choice of the value for μ is critical to the success of relaxation and the usual range is $0 < \mu < 2$ with the two cases distinguished as under-relaxation ($\mu < 1$) and over-relaxation ($\mu > 1$). When this method is applied successively, then it is referred to as successive over-relaxation, or by the mnemonic SOR.

Although some theoretical work has been done in estimating the exact choice of the constant μ, it is usually treated as an empirical parameter to be determined for each different problem. The relaxation method finds an important application in large systems of equations that have to be solved repeatedly for different, but similar, matrices. However, stability and convergence characteristics are issues that often come to the fore when large rank systems of equations are to be solved.

This chapter is concluded with Tables 5.9 and 5.10 that are guides to numbered equations corresponding to the solution methods for systems of linear equations described here.

EXAMPLES 5.30–5.32

5.12 Sparse Matrices: Case Study

EXAMPLE 5.4

Table 5.9 Direct methods

Method	Equations
Gauss elimination	5.79 to 5.85
Partial pivoting	5.96
Doolittle	Table 5.6, 5.113 to 5.121
Crout factorization	Table 5.7, 5.122 to 5.125
Cholesky factorization	Table 5.8 and following note
Tridiagonal systems	5.129 to 5.145
Gauss-Jordan elimination	5.155 to 5.160

Table 5.10 Iterative methods $x^{<m>} = Tx^{<m-1>} + t$

Method	Equations	A	T	t
Jacobi	5.192 to 5.200	L+D+U	$-D^{-1}(L+U)$	$D^{-1}y$
Gauss-Seidel	5.204 to 5.207	L+D+U	$-(L+D)^{-1}U$	$(L+D)^{-1}y$
General	5.208 to 5.211	S+A-S	$-S^{-1}(A-S)$	$S^{-1}y$

FORTRAN CODE FPARSE

5.13 Exercises

1. For the vectors $x = y = \begin{bmatrix} 1 \\ 2 \\ 3 \end{bmatrix}$ compute the products $x \times y^T$ and $x^T \times y$ and state in each case if the result is a scalar, vector or matrix.

2. Show that the matrices $A = \begin{bmatrix} 1 & 1 & -1 \\ 1 & 2 & -2 \\ -2 & 1 & 1 \end{bmatrix}$ and $B = \begin{bmatrix} 2 & -1 & 0 \\ 1.5 & -0.5 & 0.5 \\ 2.5 & -1.5 & 0.5 \end{bmatrix}$ are inverses of each other and compute the determinants $|A|$, $|B|$, and $|AB|$.

3. If $y = \begin{bmatrix} 3 \\ 1 \\ 2 \end{bmatrix}$ solve for x in the matrix equation $Ax = y$ for the matrix A of Exercise 2.

4. For the matrices $A = \begin{bmatrix} 1 & 1 & -1 \\ 1 & 2 & -2 \\ -2 & 1 & 1 \end{bmatrix}$, $B = \begin{bmatrix} 1 & 1 & 1 \\ 3 & 2 & 2 \\ 2 & 1 & 1 \end{bmatrix}$ compute

   (a) the product AB,
   (b) the product $B^T A^T$,
   (c) the result $(AB)^T$.

5. For the matrix $A = \begin{bmatrix} 4 & 1 & 1 \\ 1 & 3 & 2 \\ 1 & 2 & 2 \end{bmatrix}$ apply the vector $x = \begin{bmatrix} 1 \\ 1 \\ 1 \end{bmatrix}$ to compute the value of the product $x^T A x$ and hence determine if A is positive definite.

6. Show that the matrix $A = \begin{bmatrix} 1 & 2 & 1 \\ 2 & 2 & 3 \\ -1 & -3 & 0 \end{bmatrix}$ has the decomposition $A = LU$ where

   $L = \begin{bmatrix} 1 & 0 & 0 \\ 2 & 1 & 0 \\ -1 & 0.5 & 1 \end{bmatrix}$ and $U = \begin{bmatrix} 1 & 2 & 1 \\ 0 & -2 & 1 \\ 0 & 0 & 0.5 \end{bmatrix}$. Hence compute the determinants $|A|$, $|L|$ and thereby show that $|A| = |L| \times |U|$.

7. Apply Gaussian elimination (without pivoting) to solve the linear system $Ax = b$ when $A = \begin{bmatrix} 3 & 2 & 7 \\ 2 & 3 & 1 \\ 3 & 4 & 1 \end{bmatrix}$ and $b = \begin{bmatrix} 4 \\ 5 \\ 7 \end{bmatrix}$.

8. Compute the LU factorization of the matrix A in Exercise 7 by Doolittle's method.

9. Compute the inverse of the matrix $A = \begin{bmatrix} 2 & 1 & 1 \\ 1 & 2 & 1 \\ 1 & 1 & 1 \end{bmatrix}$.

5.13 Exercises

10. Use Doolittle's method to solve the tridiagonal matrix system $Ax = b$ when
$$A = \begin{bmatrix} 2 & 1 & 0 & 0 & 0 \\ 1 & 2 & 1 & 0 & 0 \\ 0 & 1 & 2 & 1 & 0 \\ 0 & 0 & 1 & 2 & 1 \\ 0 & 0 & 0 & 1 & 2 \end{bmatrix} \text{ and } b = \begin{bmatrix} 1 \\ 1 \\ 1 \\ 1 \\ 1 \end{bmatrix}.$$

11. The Jacobi iteration $x^{<m>} = Tx^{<m-1>} + t$ for a 4 × 4 matrix produces these iterates:

$m=1$	$m=2$	$m=3$	$m=4$	$m=5$
0.0	0.6000	1.0473	0.9326	1.0152
0.0	2.2727	1.7159	2.0533	1.9537
0.0	-1.1000	-0.8052	-1.0493	-0.9681
0.0	1.8750	0.8852	1.1309	0.9738

Compute the ratio $||x^{<m-1>} - x^{<m>}||_\infty / ||x^{<m>}||_\infty$ and decide if the iterates are converging.

12. Compute the norm $||T||_\infty$ for this matrix in Exercise 11
$$T = \begin{bmatrix} 0.0 & 0.1000 & -0.2000 & 0.0 \\ 0.0909 & 0.0 & 0.0909 & -0.2727 \\ -0.2000 & 0.1000 & 0.0 & 0.1000 \\ 0.0 & -0.3750 & 0.1250 & 0.0 \end{bmatrix}$$
and if $||T||_\infty \leq \tau$ apply the bound $||x^{<0>} - x^{<m>}|| \leq \frac{\tau^{m-1}}{1-\tau} ||x^{<2>} - x^{<1>}||$ to determine the minimum m value required to ensure $||x^{<0>} - x^{<m>}|| \leq 10^{-4}$ if $x^{<0>}$ is the exact solution.

13. For the 2 × 2 linear system $\begin{bmatrix} 2 & -1 \\ -1 & 2 \end{bmatrix} \begin{bmatrix} x_1 \\ x_2 \end{bmatrix} = \begin{bmatrix} 1 \\ 1 \end{bmatrix}$ the Jacobi iteration $x^{<m>} = Tx^{<m-1>} + t$ has an initial guess $\begin{bmatrix} 0.0 \\ 0.0 \end{bmatrix}$ and first iterate $\begin{bmatrix} 0.5 \\ 0.5 \end{bmatrix}$. Compute successive iterates up to $m = 5$ and from the ratio $||x^{<m-1>} - x^{<m>}||_\infty / ||x^{<m>}||_\infty$ determine if the iterates are converging.

14. For the linear system in Exercise 13 the Gauss-Seidel iteration has the matrix $T = \begin{bmatrix} 0.0 & 0.5 \\ 0.0 & 0.75 \end{bmatrix}$ and the first two iterates are $\begin{bmatrix} 0.0 \\ 0.0 \end{bmatrix}$ and $\begin{bmatrix} 0.5 \\ 0.75 \end{bmatrix}$. Compute the norm $||T||_\infty$ and then, for $m = 3, 4, 5$, apply the result $||T||_\infty \leq \tau$ in the inequality $||x^{<0>} - x^{<m>}|| \leq \frac{\tau^{m-1}}{1-\tau} ||x^{<2>} - x^{<1>}||$ to estimate a bound on the distance of the corresponding iterate from the exact result. Compare these bounds to those for the Jacobi case of Exercise 13. Which of these two methods appears to have the faster convergence rate?

5.14 Programming Problems

Except for trivial problems like those in the EXAMPLES and Exercises, readers are not encouraged to write their own code for linear algebra or eigenvalue problems. Instead it is strongly recommended that they familiarize themselves with the resources listed here that are available for download at no charge. A more complete list may be found on the internet by searching for the key words "linear algebra libraries". If you have access to a High Performance Computer, or own a vendor supported compiler for Fortran or C, then vendor proprietary libraries should be available to you. Otherwise try one of the ones listed here where documentation is also available.

1. BLAS (Basic Linear Algebra Subprograms) is an API standard for publishing libraries to perform basic linear algebra operations such as vector and matrix multiplication and is a Fortran library with Fortran 95, Fortran 77 and C interfaces. They are routines that provide standard building blocks for performing basic vector and matrix operations. Each routine has a name which specifies the operation, the type of matrices involved and their precision.

   (a) Visit the BLAS Technical Forum at
   https://www.netlib.org/blas/blast-forum/ for documentation and download information.
   (b) If you use the GNU Fortran compiler you could investigate what the GNU Scientific Library includes at
   https://www.gnu.org/software/gsl/doc/html/.

2. LINPACK, originally written for Fortran, is a library of linear algebra functions and subroutines. It was created in the 1970s by Jim Bunch, Jack Dongarra, Cleve Moler, and Gilbert Stewart. Intended for use on supercomputers, LINPACK solves systems of linear equations using matrices and numerical methods. Examples of included subroutines are for Gaussian elimination, finding the determinant/inverse of a square matrix, and transformations into diagonal matrices. These algorithms use column-based operations to take advantage of locality of reference.

   (a) LINPACK for Fortran has information at
   https://www.netlib.org/linpack/.
   (b) LINPACK uses BLAS routines.
   (c) LINPACK has been used as a benchmark for HPC platforms as described here https://www.top500.org/project/linpack/.

3. LAPACK (Linear Algebra PACKage) is a Fortran library of linear algebra functions. It was originally written in 1992 to replace LINPACK and EISPACK in modern vector computers with shared memory at that time. This is done by limiting the amount of data movement in using block matrix operations such as matrix multiplication. Similarly to LINPACK it was written to solve simultaneous linear equations, least-squares solutions of linear systems of equations, eigenvalue prob-

lems, and singular value problems. It has since been improved upon for distributed memory systems in ScaLAPACK.

(a) The LAPACK download site is https://www.netlib.org/lapack/.
(b) LAPACK uses BLAS routines.
(c) LAPACK is freely available but readers are advised to study the legal and licensing restrictions at the download site.
(d) LAPACK has a C language interface.
(e) LAPACK has a build under MS Windows©
(f) A User guide [35] is available at https://epubs.siam.org/doi/book/10.1137/1.9780898719604.

References

1. Kolmogorov AN, Fomin SV (1957) Elements of the theory of functions and functional analysis, vol I. Graylock Press, Rochester, NY
2. Delic G, Janse Van Rensburg EJ, Welke G (1987) A non-self-adjoint general matrix eigenvalue problem. J Comput Phys 69:325–340
3. Flügge S (1984) Practical quantum mechanics. Springer-Verlag, Berlin
4. Gantmacher FR (1977) The theory of matrices, vol I. Chelsea, New York
5. Duff IS, Erisman AM, Reid JK (2017) Direct methods for sparse matrices. Oxford University Press, Oxford, UK
6. Davis TA (2006) Direct methods for sparse linear systems. Society for Industrial and Applied Mathematics, Philadelphia, PA
7. Björck Å (1996) Numerical methods for least squares problems. Society for Industrial and Applied Mathematics, Philadelphia, PA
8. Delic G (2019) A thread parallel sparse matrix chemistry algorithm for the community multi-scale air quality model. Mod Environ Sci Eng 5:775–791
9. Delic G, Cash GG (2000) The permanent of 0,1 matrices and kallman's algorithm. Comput Phys Commun 124:315–329
10. Alan G et al (eds) (1993) Graph theory and sparse matrix computations, vol 56. The IMA volumes in mathematics and its applications. Springer-Verlag, New York
11. Deif AS (1982) Advanced matrix theory for scientists and engineers. Abacas Press, London, UK
12. George A, Liu JW-H (1981) Computer solution of large sparse positive definite systems. Prentice-Hall series in computational mathematics. Prentice-Hall Inc, Englewood Cliffs, NJ
13. Parlett BN (1980) The symmetric eigenvalue problem. Prentice-Hall series in computational mathematics. Prentice-Hall Inc, Englewood Cliffs, NJ
14. Aitken AC (1967) Determinants and matrices. Oliver and Boyd, Edinburgh, UK
15. Cullen CG (1967) Matrices and linear transformations. Addison-Wesley series in mathematics. Addison-Wesley Publishing Company, Reading, MA
16. Golub GH, Van Loan CF (1989) Matrix computations, 2nd edn. Johns Hopkins series in mathematical sciences. The Johns Hopkins University Press, Baltimore, MD
17. Muir T (1960) A treatise on the theory of determinants. Dover Publications, New York, NY
18. Householder AS (1964) The theory of matrices in numerical analysis. Dover Publications, New York, NY

19. Trefethen LN, Bau III D (1997) Numerical linear algebra. Society for Industrial and Applied Mathematics, Philadelphia, PA
20. Demmel JW (1997) Applied numerical linear algebra. Society for Industrial and Applied Mathematics, Philadelphia, PA
21. Gourlay AR, Watson GA (1973) Computational methods for matrix eigenproblems. John Wiley and Sons, New York, NY
22. Faddeeva VN (1959) Computational methods of linear algebra. Dover Publications, New York, NY
23. Lehoucq RB, Sorensen DC, Yang C (1998) ARPACK users' guide. Society for Industrial and Applied Mathematics, Philadelphia, PA
24. Barrett R et al (1994) Templates for the solution of linear systems: building blocks for iterative methods. Society for Industrial and Applied Mathematics, Philadelphia, PA
25. Smith BT et al (1976) Matrix eigensystem routines - EISPACK guide, vol 6. Lecture notes in computer science. Springer-Verlag, Berlin
26. Garbow BS et al (1977) Matrix eigensystem routines - EISPACK guide extension, vol 51. Lecture notes in computer science. Springer-Verlag, Berlin
27. Dongarra JJ et al (1998) Numerical linear algebra for high performance computers. Society for Industrial and Applied Mathematics, Philadelphia, PA
28. Dongarra JJ et al (1993) Solving linear systems on vector and shared memory computers. Society for Industrial and Applied Mathematics, Philadelphia, PA
29. Stewart GW (1998) Matrix algorithms, vol I: basic decompositions. Society for Industrial and Applied Mathematics, Philadelphia, PA
30. Gallivan KA et al (1990) Parallel algorithms for matrix computations. Society for Industrial and Applied Mathematics, Philadelphia, PA
31. Coleman TF, Van Loan C (1988) Handbook for matrix computations. Society for Industrial and Applied Mathematics, Philadelphia, PA
32. Golub GH, Van Loan CF (2013) Matrix computations, 4th edn. Johns Hopkins University Press, Baltimore, MD
33. Osterby O, Zlatev Z (1983) Direct methods for sparse matrices. Springer-Verlag, Berlin
34. Higham NJ (1996) Accuracy and stability of numerical algorithms. Society for Industrial and Applied Mathematics, Philadelphia, PA
35. Anderson E et al (1999) LAPACK users' guide, 3rd edn. Society for Industrial and Applied Mathematics, Philadelphia, PA

Finding Roots of Functions 6

6.1 Definition and Examples

A root of a function of one variable is the value of the argument for which the function has a zero value. The roots of some of the functions introduced in Chap. 2 play an important role in subsequent chapters and these are explored in examples and case studies. Simple examples are the polynomials in Table 2.2. The Chebyshev polynomials $T_n(x)$ have roots whose locations are known, whereas the Legendre polynomials $P_n(x)$ generally have roots that are not known analytically and need to be found by numerical algorithms such as those described in this chapter.

If there is more than one zero value of the function, then there is more than one root if they are distinct. For example, both $T_1(x)$ and $P_1(x)$ have only one root at $x = 0$, whereas $T_2(x)$ and $P_2(x)$ each have two distinct roots

$$P_2(x): \ x_1 = -\sqrt{\frac{1}{3}}, \ x_2 = +\sqrt{\frac{1}{3}}, \tag{6.1}$$

$$T_2(x): \ x_1 = -\sqrt{\frac{1}{2}}, \ x_2 = +\sqrt{\frac{1}{2}}. \tag{6.2}$$

Note that the number of roots in the case of polynomials is often equal to the highest power of the argument. Thus $P_5(x)$, $T_5(x)$ each have five roots, one of which is at the origin and the others are symmetric pairs with respect to the origin. The roots of higher order polynomials may be obtained by factorization, but in the case of a function that is not a simple polynomial, finding the roots algebraically is not the simplest method. In such cases the easiest approach is a numerical algorithm.

EXAMPLES 6.1–6.2

Supplementary Information The online version contains supplementary material available at https://doi.org/10.1007/978-3-031-90178-2_6

6.2 Numerical Root Finding Methods

A numerical method of finding the root of a function, as opposed to an algebraic one, is based on the assumption that the value of the root is not required to infinite precision and that an approximation to it will suffice. A numerical method may therefore proceed by constructing a local polynomial approximation and using the root of this polynomial. Often simple geometrical ideas are employed and the estimate for the root is successively improved upon until the approximate value is within some prescribed distance from the exact value. However, convergence to the exact value of the root is not always guaranteed and needs closer study. Three simple numerical root finding methods are discussed here and each uses some information concerning the slope of the objective function. The methods are generally useful because they require no special conditions on the objective function, except that it is continuous and smooth. The smoothness requirement implies the existence of at least one derivative.

EXAMPLE 6.3

6.3 The Bisection Method

The bisection method uses a simple geometric concept: choose a finite interval of the argument that is known to straddle the location of the root. This method bisects the interval and retains the part of it on which the function changes sign. Repeated application of the method will give convergence, if it is assumed that the function has only one root on the interval in question.

Before describing the bisection algorithm, consider the implication of the intermediate value theorem (see p95 of [1]) for a function that is a continuous function of its argument. For a function $f(x)$ defined on an interval $x \in [a, b]$, if either $f(a) \geq y \geq f(b)$, or $f(a) \leq y \leq f(b)$, then $c \in [a, b]$ exists such that $f(c) = y$. A special case occurs for $y = 0$, when either $f(a) \leq 0, f(b) \geq 0$, or $f(a) \geq 0, f(b) \leq 0$, and in either case c is a root. Obviously, the product $f(a)f(b)$ is negative when the interval $[a, b]$ straddles the root and, when the product is not negative, it does not straddle a root.

The bisection algorithm then consists of the following steps:

1. $c = (a + b)/2$;
2. $u = f(a), v = f(b), w = f(c)$;
3. if $wu < 0$ then store $w \to v$ and $c \to b$, otherwise;
4. if $wv < 0$ then store $w \to u$ and $c \to a$, otherwise;
5. if $w = 0$ then stop.

Cases 3 and 4 are shown schematically in Fig. 6.1a, b and case 5 is shown in Fig. 6.1c, while Fig. 6.1d is the case of no root in the search interval.

Step 5 corresponds to location of the root, otherwise the bisection process is continued and a sequence $a_i, b_i, c_i, i = 1, 2, \ldots$, is generated by repeated application

6.3 The Bisection Method

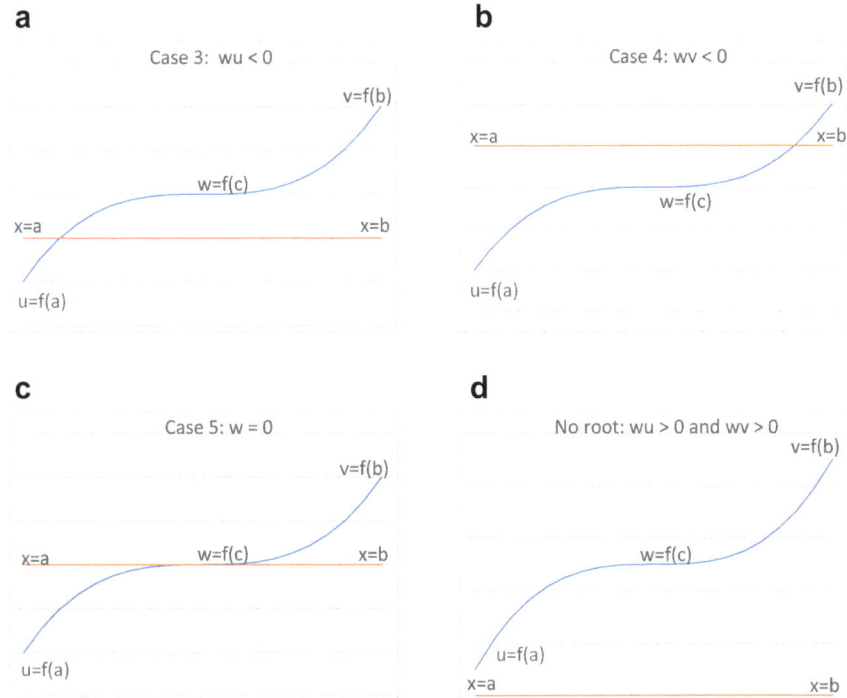

Fig. 6.1 (a) Case 3, (b) Case 4, (c) Case 5, and (d) no root

of steps 1 to 5. The iteration may be terminated either by inspection of the interval size, $|b_i - a_i|$, or the function value $|f(c)|$. If ϵ_1 and ϵ_2 are the respective tolerances, then either one may be used to terminate the iteration depending on either of these criteria

$$|b - a| < \epsilon_1, \tag{6.3}$$

$$|f(c)| < \epsilon_2. \tag{6.4}$$

Clearly, the bisection method does converge to the root if the assumptions are satisfied. The assumptions required for convergence are that

- $[a, b]$ straddles the root,
- there is only one root, and
- $f(x)$ is continuous on $[a, b]$.

However, what is not evident is the rate of convergence, or the rate with which c_i approaches the exact value of the root x_r as a function of the number of iterations. To investigate this let $|x_r - c_0|$ be the initial distance from the (unknown) root x_r, as shown in Fig. 6.2.

$$a_0 \mid \!\!-\!\!-\!\!-\!\!-\!\!-\!\!-\!\!-\!\!-x_r<\!\!-\!\!-\!\!-\!\!-\!\!-\mid x_r - c_0\mid\!\!-\!\!-\!\!-\!\!-\!\!->c_0\!\!-\!\!-\!\!-\!\!-\!\!-\!\!-\!\!-\mid b_0$$

Fig. 6.2 Location of root and distance from it

From Fig. 6.2 it follows that

$$\frac{1}{2}|b_0 - a_0| \geq |x_r - c_0|, \tag{6.5}$$

and in the subsequent iterations similar bounds may be established as

$$\frac{1}{4}|b_0 - a_0| \geq \frac{1}{2}|b_1 - a_1| \geq |x_r - c_1|, \tag{6.6}$$

$$\frac{1}{2^{n+1}}|b_0 - a_0| \geq \frac{1}{2}|b_n - a_n| \geq |x_r - c_n|. \tag{6.7}$$

In view of the last result it is possible to make a statement on the rate of convergence of the bisection method. For $f(x)$ continuous on $x \in [a, b]$ and $f(a)f(b) \leq 0$, then the error in the root for the bisection method with n iterations will not exceed $|b-a|/2^{n+1}$. Conversely, the result of (6.7) may be used to estimate the number of iterations required to ensure that an error $\epsilon \approx |x_r - c_n|$ is reached. The result follows from the logarithm of both sides of (6.7)

$$n > \frac{log(b-a) - log(2\varepsilon)}{log(2)}. \tag{6.8}$$

EXAMPLES 6.4–6.8

6.4 The Newton Method

Newton's algorithm uses information concerning the slope of the objective function to determine the direction in which the iteration should progress. It does this by application of a local linear approximation and constructs a tangent whose intercept on the abscissa provides an estimate of the root. In view of the importance and utility of the method in diverse applications, two derivations are presented here. The first is a heuristic derivation using a geometrical construction and the second is based on Taylor's series (2.19).

Consider Fig. 6.3 where $\ell_1(x)$ is the tangent to $f(x)$ at the point $x = x_0$ and $\ell_1(x)$ has the intercept x_1 with the x-axis while the actual location of the root is at x_r. On applying a change of variable $z = x - x_0$, that is equivalent to moving the origin to $x = x_0$, the equation of the straight line ℓ_1 is given by

$$\ell_1 = \text{slope} \times z + \text{intercept on the ordinate}, \tag{6.9}$$

6.4 The Newton Method

Fig. 6.3 Tangent and intercept

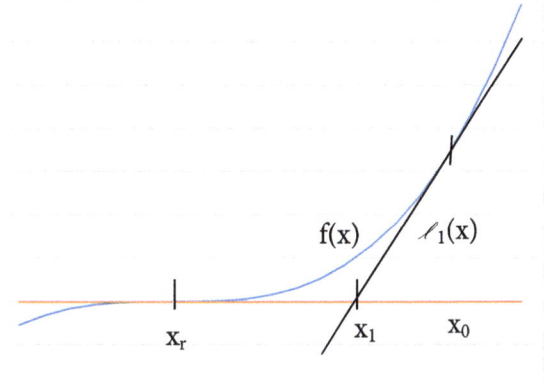

or

$$\ell_1(x) = f'(x_0)[x - x_0] + f(x_0), \tag{6.10}$$

where, at $x = x_0$, $\ell_1(x_0) = f(x_0)$. The estimate for the root of $f(x)$ is then the root of $\ell_1(x)$ which follows from (6.10)

$$0 = f'(x_0)[x_1 - x_0] + f(x_0), \tag{6.11}$$

and

$$x_1 = x_0 - \frac{f(x_0)}{f'(x_0)}. \tag{6.12}$$

At the point x_1 a second tangent, $\ell_2(x)$, is constructed and the intercept on the x-axis yields a second estimate, x_2, of the root. Continued application of this method generates a sequence

$$x_{i+1} = x_i - \frac{f(x_i)}{f'(x_i)}, \quad i = 0, 1, 2, \ldots. \tag{6.13}$$

Another derivation of Newton's method follows on seeking the solution h that solves $f(x_0 + h) = 0$ after application of the Taylor series of (2.32)

$$f(x_0 + h) = f(x_0) + hf^{(1)}(x_0) + \frac{1}{2}h^2 f^{(2)}(x_0) + \cdots \tag{6.14}$$

where it should be noted that in this expression x_0 is a constant and h is the variable. If it is assumed that terms of order h^2 and higher are negligibly small for sufficiently small h, then solution for h in

$$f(x_0 + h) \approx f(x_0) + hf^{(1)}(x_0),$$
$$= 0 \tag{6.15}$$

yields
$$h = -\frac{f(x_0)}{f^{(1)}(x_0)}, \quad (6.16)$$

and $x_1 = x_0 + h$ as in (6.12).

While the Newton method generates a sequence of approximants $x_0, x_1, x_2, \ldots$, it is meaningful to ask under which conditions does the sequence converge to a root x_r and what is the rate of convergence? Applying (2.31) for the average value of the first derivative between x_i and x_r

$$f(x_i) - f(x_r) = f^{(1)}(\xi_i)[x_i - x_r], \quad (6.17)$$

then, from the result $f(x_r) = 0$, it follows that

$$[x_r - x_i] = -\frac{f(x_i)}{f^{(1)}(\xi_i)},$$
$$\approx -\frac{f(x_i)}{f^{(1)}(x_i)},$$
$$= x_{i+1} - x_i. \quad (6.18)$$

In deriving (6.18) it is assumed that when x_i is close to x_r, the average value of the derivative $f^{(1)}(\xi_i)$ is very close to $f^{(1)}(x_i)$ on the interval $[x_r, x_i]$.

To investigate convergence, (6.18) is used after an application of Taylor's theorem at a root x_r:

$$f(x_r) = f(x_i) + f^{(1)}(x_i)[x_r - x_i] + \frac{1}{2}f^{(2)}(\xi_i)[x_r - x_i]^2, \quad (6.19)$$

where the last term is the Lagrange form of the remainder in Newton's method (2.28). The left-hand side is zero and division by $f^{(1)}(x_i)$ gives

$$0 = \frac{f(x_i)}{f^{(1)}(x_i)} + [x_r - x_i] + \frac{1}{2}[x_r - x_i]^2 \frac{f^{(2)}(\xi_i)}{f^{(1)}(x_i)}. \quad (6.20)$$

On substitution for the first term by (6.18) the last expression yields

$$[x_r - x_{i+1}] = -\frac{1}{2}[x_r - x_i]^2 \frac{f^{(2)}(\xi_i)}{f^{(1)}(x_i)}, \quad (6.21)$$

and again the near equivalence of $f^{(1)}(\xi_i)$ and $f^{(1)}(x_i)$ means that the ratio of the two derivatives in (6.21) may be treated as a constant on the interval $[x_r, x_i]$. Then substitution of the constant C defined as

$$C = -\frac{f^{(2)}(\xi_i)}{2f^{(1)}(x_i)}, \quad (6.22)$$

into the result of (6.21) establishes an important inequality

$$|x_r - x_{i+1}| \leq |C| \times |x_r - x_i|^2, \quad (6.23)$$

and for x_i close to x_r, the iterations have quadratic convergence to x_r. To demonstrate the importance of this result assume, $C = 1$ and that the ith iterate has converged to within one digit in the kth decimal place, or

$$|x_r - x_i| \le 10^{-k}, \tag{6.24}$$

then, from (6.23), quadratic convergence means that

$$|x_r - x_{i+1}| \le 10^{-2k}. \tag{6.25}$$

Therefore, the $(i + 1)$st iterate has converged to within one digit in the $2k$th decimal place and the number of significant figures in the approximation of the root has doubled between the ith and $(i + 1)$st iterate. This is a very powerful result but it is subject to the assumption that the initial iterate x_0 is close to the root x_r and that the constant C is not large. In particular, convergence would require that $|C(x_r - x_0)| < 1$, or

$$|x_r - x_0| \le \frac{1}{|C|}, \tag{6.26}$$

therefore a bound on how far x_0 should be from the root could be determined by computing the constant C from (6.22). However, this is not usually done in view of the expense of computing estimates of the first two derivatives of the function $f(x)$.

EXAMPLES 6.9–6.13

6.5 The Secant Method

Newton's method requires an analytical expression for the derivative of the function whose root is to be found. If the function has a complicated form, then the analytical expression for the derivative may be difficult, or impossible to obtain. In such a case the derivative may be estimated numerically which is the case in the secant algorithm.

From calculus, the derivative of $f(x)$ at x is defined as

$$f'(x) = \lim_{h \to 0} \frac{f(x+h) - f(x)}{h}, \tag{6.27}$$

and a suitable numerical approximation for small h is to use the right-hand side of (6.27)

$$f'(x_i) \approx \frac{f(x_{i-1}) - f(x_i)}{x_{i-1} - x_i}. \tag{6.28}$$

As shown in Fig. 6.4, (6.28) is the slope of a secant to $f(x)$ between x_i and x_{i-1} and that is the origin of the method's name.

Thus, given two starting values x_0 and x_1, the secant method of locating a root generates a sequence of approximations from the iteration

$$x_{i+1} = x_i - f(x_i) \left(\frac{x_i - x_{i-1}}{f(x_i) - f(x_{i-1})} \right). \tag{6.29}$$

Fig. 6.4 Secant and intercept

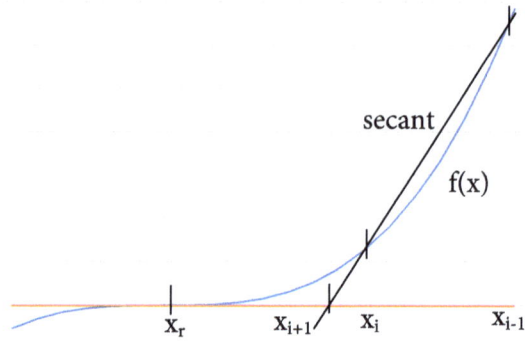

The secant method requires both the current and previous function value to generate the next iteration and convergence of the sequence would imply that in the denominator, $f(x_{i-1}) \to f(x_i)$, which would lead to loss of numerical significance in (6.29). Therefore, with a suitable choice of ϵ, an appropriate termination criterion could be

$$|f(x_i) - f(x_{i-1})| \leq \epsilon |f(x_i)|. \tag{6.30}$$

However, the rate of convergence of the secant method is determined by the relation:

$$|x_r - x_{i+1}| \leq |C'| \times |x_r - x_i|^\alpha, \tag{6.31}$$

where C' is a constant dependent on the function $f(x)$ and the convergence factor is

$$\alpha = \frac{1}{2}(1 + \sqrt{5}) \approx 1.62. \tag{6.32}$$

Thus the exponent factor has the bounds $1 < \alpha < 2$ and convergence of the secant method is said to be super-linear (above linear), but obviously below quadratic. For a discussion of error analysis and derivation of the values for α, see Chap. 4 of [2].

EXAMPLES 6.14–6.17

6.6 Comparison of Methods

Table 6.1 summarizes attributes of the three root finding methods described here. Each method has advantages and disadvantages and the one chosen depends on the properties of $f(x)$ in the region of x to which the method is applied. The bisection method will always converge to the root whenever the root is straddled, but convergence is always slow. On the other hand, Newton's method, while it often converges rapidly, may give problems in cases when the derivative is small but the function

Table 6.1 Root finding methods

Property	Bisection	Secant	Newton
Exponent	1	1.62	2
Rate	Slowest		Fastest
Computation	$f(x_i)$	$f(x_i)$	$f(x_i)$ and $f'(x_i)$
Requirements	$f(x_i)$ and $f(x_{i-1})$	$f(x_i) - f(x_{i-1})$	$f'(x_i)$
	Must straddle root	Not be too small	Not zero near root

value is large, as is the case close to extrema. Conversely, the secant method can develop problems with the denominator when approaching the root.

It is obvious that care should always be exercised in applying any of these root finding methods and a well-designed algorithm should test for known problems in each method. Fortunately, in root finding the simplest criterion is also the most secure one, namely, if the algorithm is converging, then both the function value and the difference between successive iterates must approach zero, even if they do so at different rates.

6.7 Finding Roots of Ill-Posed Problems

Each function is, of course, different, but even if it is assumed that smoothness holds and no discontinuity is present, other difficulties may still arise in addition to the problems already detailed above. Two simple examples of possible pathologies are discussed here to show how even a method with good convergence properties may behave poorly.

The first example of a problematic case is that of a multiple root. A root x_r, of a function $f(x)$, is said to have a multiplicity m if it is found that

$$f(x_r) = f'(x_r) = \cdots f^{(m-1)}(x_r) = 0, \quad (6.33)$$

and

$$f^{(m)}(x_r) \neq 0. \quad (6.34)$$

A special case occurs if the function $f(x)$ has the form

$$f(x) = (x - x_r)^m u(x), \quad (6.35)$$

where $u(x_r) \neq 0$, and $u(x)$ is also a smooth function. Direct application of Newton's method to a function with a multiple root will result in poor convergence that becomes worse with increasing multiplicity. One approach to circumvent the problem is simply to successively differentiate the function and search for the root of each derivative until sufficiently rapid convergence is obtained.

The second example of problems related to the properties of the objective function is that of stability. In a function where the location of a root x_r is a function of some parameter β, then, under normal circumstances, the location of the root $x_r(\beta)$ changes smoothly as β varies. However, if this is not the case and small changes in the parameter β gives rapid, or erratic, movement of the root location on the real line, or even movement into the complex plane, then the root finding problem is unstable because of an instability in the function itself. The root finding problem may also be unstable numerically because small errors in the function value produce large errors in the value of the root. In such cases the problem is said to be ill-posed or ill-conditioned.

EXAMPLES 6.18–6.20

6.8 Roots of the Spherical Bessel Function: Case Study

The spherical Bessel function was introduced in Sect. 2.1 and defined in (2.98) of Sect. 2.6. It is defined on the interval $[0, \infty]$ and has many roots. Their values may be found in tables or, alternatively, by the root finding methods described in the preceding sections. The roots of the spherical Bessel function interlace as their order increases and this property may be used to provide bounds on the roots of the next higher order. For example, the roots of the spherical Bessel function of order zero are multiples of π and provide bounds on the roots of the order one function. Once these are found, they may be used as bounds for the order two function and so on. The examples show some cases and the bisection method is applied using these bounds. Once a few iterations of the bisection method establish the locality of a root, or if the convergence is too slow, a switch to the Newton method would accelerate convergence to the root.

EXAMPLES 6.21–6.24

6.9 Roots of the Legendre Polynomial: Case Study

For a second case study, two Fortran codes may be applied to find the roots of Legendre polynomials up to order of 150 and the weights and abscissas for Gauss-Legendre quadrature discussed in Sect. 8.2.2. The accuracy and precision of the Fortran codes presented here extend to beyond double precision and care must be exercised that the host platform has sufficient word length to support this demand.

FORTRAN CODES PLZEROES

6.10 Exercises

1. Find all roots of the Legendre polynomial $P_6(x)$ to three significant figures by noting the number of iterations with

   (a) the bisection method,
   (b) Newton's method,
   (c) the secant method,
   (d) compare these three methods and decide which has the lowest number of function evaluations.

2. Apply the bisection, Newton and secant methods to solve for the roots of the following functions to six significant figures. Tabulate the iteration number and values of x and $f(x)$ at each iterate and state your observations when comparing convergence rates of each method.

   (a) $f(x) = 2x^6 - x - 1.5$,
   (b) $f(x) = xe^x - 1$,
   (c) $f(x) = 0.01x^3 - sin(x)$.

3. Use Newton's method to find the first negative root of the following functions to six significant figures. Tabulate the iteration number and values of x and $f(x)$ at each iterate and state the observations you see when comparing convergence rates of each function.

   (a) $f(x) = -0.0115x^3 - sin(x)$,
   (b) $f(x) = -0.015x^3 - sin(x)$.

4. Use Newton's method to find the first positive root of $f(x) = x^5 - 4cos(x)$ using $x_0 = 0.9$ as the staring value. Show a table of iterate values for x_i, $f(x_i)$ with the number of iterations required for four significant figures in $f(x)$.

5. Use Newton's method to find the first positive root of $f(x) = \frac{sin(x)}{x} - cos(x)$ using $x_0 = 4$ as the staring value. Show a table of iterate values for x_i, $f(x_i)$ with the number of iterations required for five significant figures in $f(x)$.

6.11 Programming Problems

1. Write a code to apply the bisection method to locate the root of a function to a specified accuracy of ε. Test the code with a simple known function such as $f(x) = P_6(x)$.
2. Apply the program to locate the zeros x_{jn} of the Legendre polynomial $P_n(x)$ of order n indexed such that $-1 < x_{1n} < x_{2n} < \cdots < x_{n-1\,n} < x_{nn} < +1$ for n up to order $n \le 80$, or the largest order possible for the available software

and hardware combination. To evaluate the function values, use the recurrence relation (2.43) starting with $P_0(x) = 1$ and $P_1(x) = x$. To obtain bounds on roots at successive orders, apply the interlacing property of the polynomial roots.

(a) Evaluate the efficiency of the method by counting the number of function evaluations for each root to reach the specified accuracy ε.
(b) Tabulate the function value at each iterate.
(c) Compare the rate of convergence of the method for a small and large value of the order n for the two roots nearest 0 and $+1$, respectively.
(d) Using the best values obtained for the roots at each order compute the values

$$w_{jn} = \frac{2(1 - x_{jn}^2)}{[nP_{n-1}(x_{jn})]^2},$$

$$W_n = \sum_{j=1}^{n} w_{jn}.$$

(e) To check the accuracy of the calculation, inspect the difference $|2 - W_n|$ and tabulate the result for each $n > 1$.

3. Repeat the Programming Problem 2 by modifying the code to use the secant method and compare results with the bisection method. Which of the two methods is more efficient and why?
4. Modify the code developed for Problems 1 to 3 to find the roots of the spherical Bessel function $j_\ell(r)$, $r \in [0, 25]$, and $\ell = 0, 1, 2$, using the algorithm of EXAMPLES 6.21 to 6.24 for bisection, Newton, and secant methods.

(a) Evaluate the efficiency of the method by counting the number of function evaluations for each root to reach the specified accuracy ε.
(b) Tabulate the function value at each iterate.
(c) Compare the rate of convergence of all three methods for orders $\ell = 0, 1, 2$, and the smallest and largest roots on the interval $[0, 25]$.

References

1. Thomas GB (1960) Calculus and analytical geometry. Addison-Wesley Publishing Company Inc, Reading, MA
2. Atkinson KE (1985) Elementary numerical analysis. John Wiley and Sons, New York

One-Dimensional Numerical Integration 7

7.1 Definition of the Integral

In calculus the integral occurs as one of two types, either as an indefinite integral

$$\int f(x)dx = F(x) + \alpha, \tag{7.1}$$

or, as a definite integral

$$\int_a^b f(x)dx = F(b) - F(a), \tag{7.2}$$

where $F(x)$ is a function whose derivative is $f(x)$, defined on the interval $x \in [a, b]$, and α, $F(b)$, $F(a)$ are constants. The result of the indefinite integral is a function whereas the result of the definite integral is a number, intuitively thought of as the area under the curve of $f(x)$. When the definite integral is required in calculus, or analysis, it is customary to first determine the unknown function $F(x)$ and then perform the substitution $F(b) - F(a)$ to determine the numerical value of the definite integral. This is usually only possible if $f(x)$ is not a complicated function and its definite integral $F(x)$ is easily determined or, at worst, obtained from known tables of integrals [1–3]. However, when such an approach is not possible, then the definite integral may be evaluated approximately by a numerical technique [4–9]. Such numerical integration methods are the subject of this chapter. Solutions obtained for the definite integral by consulting tables of integrals, or applying symbolic manipulation software, will not be discussed here and the coverage is not exhaustive.

Supplementary Information The online version contains supplementary material available at https://doi.org/10.1007/978-3-031-90178-2_7

© The Author(s), under exclusive license to Springer Nature Switzerland AG 2026
G. Delic, *Guide to Numerical Algorithm Design and Development*,
Texts in Computer Science, https://doi.org/10.1007/978-3-031-90178-2_7

Numerical integration consists of replacing the integral by a finite sum that approximates the value of the integral. Thus, if

$$\Im(f) = \int_a^b f(x)dx, \qquad (7.3)$$

then an approximation is chosen as

$$S_n(f) = \sum_{i=1}^n w_i f_i, \qquad (7.4)$$

where the function values $f_i = f(x_i)$ are computed at n discrete points and, respectively, multiplied by each of the n discrete weights w_i. The approximation (7.4) is known as a quadrature formula and values for the number, n, and weights w_i are chosen so as to bound the error of the approximation

$$E_n = |\Im(f) - S_n(f)|. \qquad (7.5)$$

The numerical integration formulas discussed in this chapter fall into two broad types. In the Newton-Cotes-type quadrature formulas, the integral is replaced by a finite sum of definite integrals over a sequence of sub-intervals. On each sub-interval, $f(x)$ is approximated as the sum of integrals of a polynomial over each sub-interval. In Newton-Cotes-type quadratures the polynomial is usually of low order and the function values $f_i = f(x_i)$ are at regularly spaced values of the independent variable x. In this way the error of (7.5) may be made small by increasing either the order of the polynomial, or the number of points n. On the other hand, in Gauss-type quadrature formulas the quadrature sum of (7.4) is chosen to exactly reproduce the value of the integral (7.3) subject to the assumption that, on the interval $[a, b]$, $f(x)$ is a polynomial of order not exceeding a specific rank (to be explained in more detail below). These two types of quadrature methods are described in more detail in following sections, after the basic definition of the definite integral. A third method is a statistical one, as shown in the examples, but it is not commonly applied for integrals in one dimension (see Sect. 8.4).

EXAMPLES 7.1–7.2

7.2 The Riemann Integral

Let $f(x)$ be defined on the closed interval $x \in [a, b]$ that is subdivided into sub-intervals $[x_i, x_{i+1}]$, i=1,..., $n-1$ ordered such that

$$a = x_1 < x_2 < \cdots < x_{n-1} < x_n = b. \qquad (7.6)$$

On each sub-interval the function $f(x)$ has both a greatest lower bound

$$\ell_i = \inf_{x \in [x_i, x_{i+1}]} f(x), \qquad (7.7)$$

7.2 The Riemann Integral

and a least upper bound

$$u_i = \sup_{x \in [x_i, x_{i+1}]} f(x). \tag{7.8}$$

Then, two sums of the type (7.4) may be defined, corresponding, respectively, to (7.7) and (7.8), as

$$S_n^L(f) = \sum_{i=1}^{n-1} \ell_i [x_{i+1} - x_i], \tag{7.9}$$

and

$$S_n^U(f) = \sum_{i=1}^{n-1} u_i [x_{i+1} - x_i] \tag{7.10}$$

that represent two estimates of the area under the graph of $f(x)$ on the interval $[a, b]$. Each term in (7.9), or (7.10), is seen to be the area of a rectangle on the sub-interval $[x_i, x_{i+1}]$ and an example for such a term is shown in Figs. 7.1 and 7.2.

It may be concluded that

$$S_n^L(f) \leq \text{area under the curve of } f(x) \leq S_n^U(f), \tag{7.11}$$

and this is true irrespective of the way in which the sub-intervals are chosen. Then, in the limit $n \to \infty$, the Riemann definition of the integral is

$$\lim_{n \to \infty} S_n^L(f) = \int_a^b f(x) dx,$$
$$= \lim_{n \to \infty} S_n^U(f), \tag{7.12}$$

and in the limit both sums (7.9) and (7.10) yield the same result. This definition of the integral assumes that $f(x)$ is continuous, or that whenever the argument x changes by a small amount, so does the function value. For discontinuous functions

Fig. 7.1 Example of a sub-interval

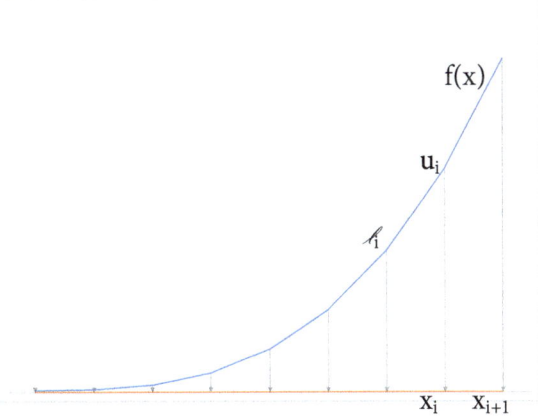

Fig. 7.2 Area calculation

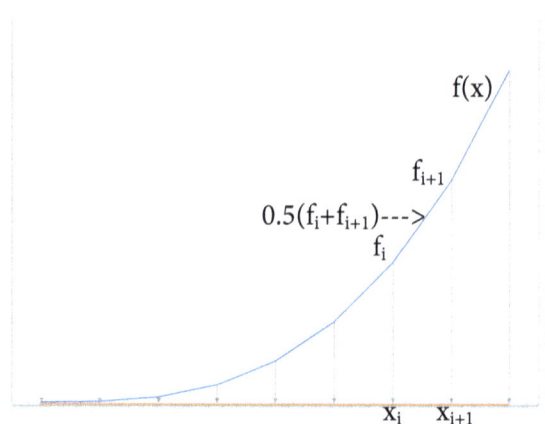

this is not the case and an alternative definition of the integral is required. However, this complication will not be explored here because only continuous functions are investigated.

The summations of (7.9) and (7.10) provide one numerical method for generating an approximation to the integral of (7.3). In view of the limiting process indicated in (7.12), it may be anticipated that by increasing n, or the number of sub-intervals, the bounds on both summations of (7.9), (7.10) would improve and, consequently, the approximation for the integral should converge. However, what is important in applications is the rate of convergence as n increases because the expense of any quadrature method is related to the number of function values, f_i, required to reach a prescribed value for the error in (7.5).

The next section explores some Newton-Cotes-type quadratures that show good convergence characteristics with increasing number of function values in the corresponding quadrature sums. These quadrature rules approximate the function $f(x)$ by a low-order polynomial on each sub-interval. The section after that introduces Gauss-type quadrature formulas that have polynomial interpolation on the entire interval and also have increasing accuracy as polynomial order increases.

EXAMPLES 7.3–7.4

7.3 Fixed Step Formulas

7.3.1 The Trapezoidal Rule

The trapezoidal rule applies the lowest order polynomial approximation to $f(x)$ on the sub-interval $[x_i, x_{i+1}]$ by simply joining the points f_i and f_{i+1} by a straight line. In place of the summations (7.9), (7.10), a single sum is used with the average value of the function on the sub-interval $\frac{1}{2}(f_i + f_{i+1})$ replacing ℓ_i or u_i. Then the

7.3 Fixed Step Formulas

contribution to the summation from each sub-interval is $\frac{1}{2}(f_i + f_{i+1}) \times [x_{i+1} - x_i]$ and the trapezoid approximation to $\Im(f)$ is the simple sum

$$S_n^T(f) = \frac{1}{2}\sum_{i=1}^{n-1}(f_i + f_{i+1})[x_{i+1} - x_i]. \tag{7.13}$$

Figure 7.2 shows the situation for the trapezoid rule on a typical sub-interval.

On the interval $[a, b]$ a special case of the trapezoid quadrature rule arises if the value of the arguments are regularly spaced as in

$$x_i = a + (i-1)h, \ 1 \le i \le n, \tag{7.14}$$

where

$$h = \frac{(b-a)}{n-1}, \tag{7.15}$$

then the expression for (7.13) simplifies to

$$S_n^T(f) = \frac{1}{2}h \sum_{i=1}^{n-1}(f_i + f_{i+1}), \tag{7.16}$$

$$= h\left[\frac{1}{2}(f_1 + f_n) + \sum_{i=2}^{n-1} f_i\right]. \tag{7.17}$$

Clearly, from the point of view of computing expense, the trapezoid estimate of (7.17) is attractive because no multiplications are required as all weights (excepting the first and last) are equal to unity.

However, the simplicity of the trapezoid quadrature is of little value without suitable convergence behavior as $n \to \infty$. For this there is a theorem on the error term of the trapezoid estimate for $\Im(f)$. This states that, given a continuous second derivative $f^{(2)}(x)$ on $[a, b]$ and a uniform step size h between grid points x_i, then

$$\Im(f) - S_n^T(f) = -\frac{(b-a)}{12}h^2 f^{(2)}(\xi) \tag{7.18}$$

for $\xi \in [a, b]$. Note that the error is $\mathcal{O}(h^2)$ and that means, if h is reduced by a factor of two, then the error is reduced by a factor of four. The proof of (7.18) is now outlined by application of the Taylor series (2.36) to $f(x)$ and $F(x)$, the definite integral of $f(x)$. This method uses the Fundamental Theorem of Calculus which states that for $F(x)$ continuous on $[a, t]$, with $f = F'$

$$\int_a^t f(x)dx = F(t), \tag{7.19}$$

from which it follows that

$$F(a) = 0. \tag{7.20}$$

Consider the sub-interval $[a, a+h]$ and apply the Taylor series (2.36)

$$F(a+h) = F(a) + hF^{(1)}(a) + \frac{h^2}{2!}F^{(2)}(a) + \frac{h^3}{3!}F^{(3)}(a) \cdots \quad (7.21)$$

that is equivalent to

$$\int_a^{a+h} f(x)dx = 0 + hf(a) + \frac{h^2}{2!}f^{(1)}(a) + \frac{h^3}{3!}f^{(2)}(a) \cdots \quad (7.22)$$

Similarly, the Taylor series for $f(a+h)$ is

$$f(a+h) = f(a) + hf^{(1)}(a) + \frac{h^2}{2!}f^{(2)}(a) + \frac{h^3}{3!}f^{(3)}(a) \cdots \quad (7.23)$$

and adding $f(a)$ to both sides and multiplying the result by $\frac{1}{2}h$ gives the expression

$$\frac{1}{2}h[f(a) + f(a+h)] = hf(a) + \frac{h^2}{2}f^{(1)}(a) + \frac{h^3}{4}f^{(2)}(a) \cdots \quad (7.24)$$

Subtracting (7.24) from (7.22) yields

$$\int_a^{a+h} f(x)dx - \frac{1}{2}h[f(a) + f(a+h)] = -\frac{h^3}{12}f^{(2)}(\xi), \quad (7.25)$$

where the second term on the left-hand side is the trapezoid rule estimate for the integral of $f(x)$ on the sub-interval $[a, a+h]$. The difference between the estimate and the exact result is seen to have a leading term on the type (7.18), because $b-a=h$ in this case. On the right-hand side of (7.25), $\xi \in [a, a+h]$, and the average value of the second derivative is used.

It is evident from the form of the error term in (7.25) that on the sub-interval $[a, a+h]$ the error is reduced by a factor of eight if the interval is subdivided, into two sub-intervals and the size of h is reduced by a factor of two. This suggests that if the process of subdivision is continued, eventually the error in the approximation of the integral will satisfy the criterion of (7.5) for a chosen tolerance E_n. To achieve this tolerance, consider the sequence obtained by subdivision in multiples of two, with repeated application of the trapezoid rule on the sub-intervals shown in Table 7.1.

It is obvious from Table 7.1 that subsequent subdivisions of the interval $[a, b]$ have some functional values f_i in common. Therefore, when the interval is subdivided only the contributions to the trapezoid sum from new function values need to be evaluated.

7.3 Fixed Step Formulas

Table 7.1 Trapezoid rule

Number of intervals	Approximant
2^0	$S_2^T(f)$
2^1	$S_3^T(f)$
2^2	$S_5^T(f)$
2^3	$S_9^T(f)$

Thus, consideration of (7.17) beginning with $2^j + 1$ grid points and increasing to $2^{j+1} + 1$ grid points shows that after subdivision, the step size is

$$h = \frac{(b-a)}{2^j}. \tag{7.26}$$

Starting with one sub-interval and $j = 0$, $2^0 + 1 = 2$, the value of the approximant is

$$S_2^T(f) = \frac{1}{2}(b-a)[f(a) + f(b)], \tag{7.27}$$

then all subsequent trapezoid approximants $j = 1, 2, 3, \ldots$ follow by only adding contributions of the new grid points

$$S_{2^{j+1}+1}^T(f) = \frac{1}{2}S_{2^j+1}^T(f) + h \sum_{i=1}^{2^{j-1}} f_{2i}. \tag{7.28}$$

Continued subdivision and repeated application of trapezoid quadratures is one method of obtaining convergence to the exact value of an integral because the process generates a convergent sequence of numbers. However, convergence acceleration techniques are then well suited to this problem and one such approach is the Richardson extrapolation method (see Sect. 3.7), also known as the Romberg algorithm.

Define the first column of a triangular array

$$\begin{bmatrix} r_{11} & 0 & 0 & \cdots & 0 \\ r_{21} & r_{22} & 0 & \cdots & \vdots \\ r_{31} & r_{32} & r_{33} & \cdots & \vdots \\ \vdots & & & \ddots & \vdots \\ r_{j1} & r_{j2} & r_{j3} & \cdots & r_{jj} \end{bmatrix}, \tag{7.29}$$

from the definition

$$r_{j1} = S_{2^j+1}^T(f), \tag{7.30}$$

then the subsequent columns are generated in turn using the recurrence

$$r_{j+1,m+1} = r_{j+1,m} + \frac{1}{4^m - 1}\left(r_{j+1,m} - r_{j,m}\right), \quad j, m \geq 1. \tag{7.31}$$

EXAMPLES 7.5–7.7

7.3.2 Simpson's Rule

The lowest order approximation to $f(x)$ on each sub-interval is the straight line joining adjacent values of $f(x)$ on the grid as shown in Fig. 7.2. The next higher order local polynomial approximation on a sub-interval is a polynomial of order two, namely, a quadratic. The definition of a local quadratic polynomial requires three grid points f_i, f_{i+1}, f_{i+2}, and for the sake of a simplified discussion, set $i = 1$ and choose

$$\begin{aligned} x_1 &= a, \\ x_2 &= a + h, \\ x_3 &= a + 2h. \end{aligned} \quad (7.32)$$

A quadratic polynomial through these three points is given by the Lagrange interpolating polynomial of (3.31)

$$p_2(x) = f_1 \ell_{21}(x) + f_2 \ell_{22}(x) + f_3 \ell_{23}(x), \quad (7.33)$$

where, in this case, the three polynomials of (3.30) reduce to

$$\begin{aligned} \ell_{21}(x) &= \frac{1}{2h^2}(x - x_2)(x - x_3), \\ \ell_{22}(x) &= -\frac{1}{h^2}(x - x_1)(x - x_3), \\ \ell_{23}(x) &= \frac{1}{2h^2}(x - x_1)(x - x_2). \end{aligned} \quad (7.34)$$

Then the next higher order approximation to $\Im(f)$ in (7.3), on this sub-interval, is

$$\int_a^{a+2h} f(x)dx \approx f_1 \int_a^{a+2h} \ell_{21}(x)dx + f_2 \int_a^{a+2h} \ell_{22}(x)dx + f_3 \int_a^{a+2h} \ell_{23}(x)dx \quad (7.35)$$

and, without loss of generality, the choice $a = 0$ allows for easy evaluation of the three integrals of Lagrange polynomials (7.34) as

$$\int_0^{2h} \ell_{21}(x)dx = \frac{h}{3},$$

$$\int_0^{2h} \ell_{22}(x)dx = \frac{4h}{3},$$

7.3 Fixed Step Formulas

Fig. 7.3 Simpson rule approximation

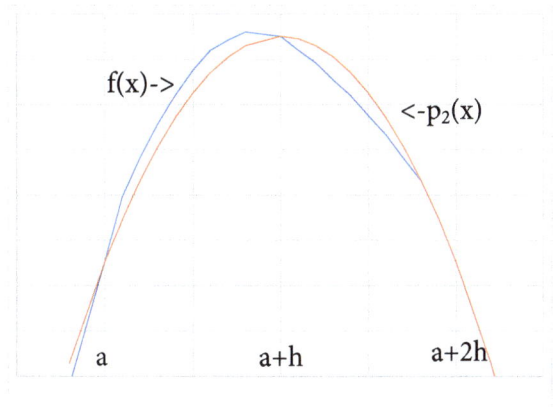

$$\int_0^{2h} \ell_{23}(x)dx = \frac{h}{3}. \qquad (7.36)$$

The weights obtained in (7.36) define Simpson's rule for which (7.4) becomes, in the case of a single sub-interval,

$$S_3^S(f) = \frac{h}{3}(f_1 + 4f_2 + f_3). \qquad (7.37)$$

The area under $f(x)$, on the sub-interval, is $\Im(f)$ and in Simpson's rule this is approximated by the area $S_3^S(f)$ under $p_2(x)$ as shown schematically in Fig. 7.3.

Whereas a sub-interval for the trapezoid quadrature rule has two grid points, in the case of Simpson's rule, it has three. To compare the two quadrature formulas, consider the simple integral

$$\int_0^{\pi} sin(x)dx = 2, \qquad (7.38)$$

for which the trapezoid approximant, with one and two sub-intervals, gives the respective values

$$S_2^T(f) = \frac{1}{2}[sin(0) + sin(\pi)], \qquad (7.39)$$

$$S_3^T(f) = \frac{1}{2}0 + \frac{1}{2}\pi \, sin(\frac{\pi}{2}), \qquad (7.40)$$

with the respective errors

$$E_2^T = 2, \qquad (7.41)$$

$$E_3^T = 0.42920367. \qquad (7.42)$$

One application of Simpson's rule gives the approximant

$$S_2^S(f) = \frac{1}{3}\frac{\pi}{2}\left[\sin(0) + 4\sin(\frac{\pi}{2}) + \sin(\pi)\right], \tag{7.43}$$

and the error is

$$E_3^S = 0.09439510. \tag{7.44}$$

From this comparison it may be concluded that for the same three grid points, one application of Simpson's rule gives an error that is a factor of 4.5 smaller than that for two applications of the trapezoid rule.

Clearly, Simpson's rule is more accurate than the trapezoidal rule, but to determine how much more accurate, some error analysis will help. As was done in Sect. 7.3.1 Taylor's theorem is applied to

$$F(a+2h) = F(a) + 2hF^{(1)}(a) + 2h^2 F^{(2)}(a) + \frac{4}{3}h^3 F^{(3)}(a) + \frac{2}{3}h^4 F^{(4)}(a) +$$

$$\frac{4}{15}h^5 F^{(5)}(a) \cdots ,$$

$$= 0 + 2hf(a) + 2h^2 f^{(1)}(a) + \frac{4}{3}h^3 f^{(2)}(a) + \frac{2}{3}h^4 f^{(3)}(a) +$$

$$\frac{4}{15}h^5 f^{(4)}(a) \cdots \tag{7.45}$$

Similarly, the Taylor series for $f(a+h)$ is

$$f(a+h) = f(a) + hf^{(1)}(a) + \frac{h^2}{2}f^{(2)}(a) + \frac{h^3}{6}f^{(3)}(a) + \frac{h^4}{24}f^{(4)}(a) \cdots \tag{7.46}$$

While that for $f(a+2h)$ is

$$f(a+2h) = f(a) + 2hf^{(1)}(a) + 2h^2 f^{(2)}(a) + \frac{4}{3}h^3 f^{(3)}(a) + \frac{2}{3}h^4 f^{(4)}(a) \cdots \tag{7.47}$$

and from these last two results it follows that

$$\frac{h}{3}[f(a) + 4f(a+h) + f(a+2h)] = 2hf(a) + 2h^2 f^{(1)}(a) + \frac{4}{3}h^3 f^{(2)}(a)$$

$$+ \frac{2}{3}h^4 f^{(3)}(a) + \frac{5}{18}h^5 f^{(4)}(a) \cdots \tag{7.48}$$

Subtracting (7.48) from (7.45) yields the result

$$\int_a^{a+2h} f(x)dx - \frac{h}{3}[f(a) + 4f(a+h) + f(a+2h)] = -\frac{1}{90}h^5 f^{(4)}(\xi), \tag{7.49}$$

where the second term on the left-hand side is Simpson's rule estimate for the integral of $f(x)$ on the sub-interval $[a, a+2h]$. The difference between this and the exact result is seen to have a leading term $\mathcal{O}(h^5)$. On the right-hand side of (7.49), $\xi \in$

7.3 Fixed Step Formulas

$[a, a + 2h]$, and the average value of the fourth derivative is used. If the value of this derivative on the interval $[a, a + 2h]$ is not large, then the error term in (7.49) is dominated by the h^5 factor.

For one sub-interval the error term of Simpson's rule in (7.49) is seen to converge at a rate $\propto h^2$ faster than that of the trapezoid rule (7.25). This behavior accounts for the large improvement in the application of Simpson's rule to the example of (7.38). To understand better the significance of the result in (7.49), consider the case that

$$f(x) = x^n, \ n < 4, \tag{7.50}$$

for which $f^{(iv)}(\xi) = 0$, and the error of Simpson's quadrature rule is zero. In other words, if the function $f(x)$ is exactly approximated by a cubic polynomial, Simpson's rule gives the exact value of the integral. For the case that $n = 4$, $f^{(iv)}(\xi) = 4!$, and the error of Simpson's quadrature rule is

$$E^S = -\frac{4}{15}h^5, \tag{7.51}$$

and each time the spacing between grid points is halved, the error is reduced by a factor of 32. This suggests that, in general, continued halving of the step size in Simpson's rule should lead to an approximant that converges rapidly to the value of the integral. Simpson's rule is special in that an error term of order h^4 could have been expected, but the analysis above showed this to be of order h^5.

EXAMPLES 7.8–7.12

7.3.3 Higher Order Rules

In Sect. 3.10, Stirling's central difference formula was used to obtain coefficients for approximants to first- and second-order derivatives of a function when the function values were available for a discrete grid of equally spaced values or the argument.

In the same way, coefficients of the approximation to the integral of the function on the same interval may also be obtained, as before, with a set of values, f_s, $s = 0, \pm 1, \pm 2, \ldots$, of a function $f(x)$ tabulated at equally spaced values of the abscissa $x = x_0 + sh$. Stirling's central difference formula (3.96) is integrated with this change in variable

$$\int_{x_{-k}}^{x_{+k}} f(x)dx = h \int_{-k}^{+k} f_s ds. \tag{7.52}$$

On the right-hand side the terms in $\mu\delta^{2k+1} f_0$ will not contribute because the integrals of their coefficients over the interval $s = -k$ to $s = +k$ are zero, while the largest difference retained will correspond to $\delta^{2k} f_0$. A $2n + 1$ point quadrature formula is obtained by setting $k = n$ for some $n = 1, 2, \ldots$. The case $n = 1$ is Simpson's rule and that for $n = 3$ is Weddle's rule. In general

$$\int_{x_{-n}}^{x_{+n}} f(x)dx = h\int_{-n}^{+n} f_s\, ds, \qquad (7.53)$$

$$= h\sum_{j=0}^{n} \alpha_j^n \sum_{i=0}^{2j} (-1)^i \binom{2j}{i} f_{j-i}, \qquad (7.54)$$

where

$$\alpha_j^n = \frac{1}{(2j)!} \int_{-n}^{+n} \alpha_j(s)\, ds, \qquad (7.55)$$

and

$$\alpha_j(s) = s^2 \left(s^2 - 1^2\right) \cdots \left[s^2 - (j-1)^2\right], \qquad (7.56)$$

with $\alpha_0(s) = 1$. Here, by analogy to (3.66), the term $\delta^{2j} f_0$ in (3.96) has been replaced by the corresponding binomial expansion for the central difference (see Sect. 25.1.2 of [10]).

This is a quadrature formula of the general form

$$\int_{-n}^{+n} f(x)dx \approx h\left[a_0^n f_0 + \sum_{j=1}^{n} a_j^n \left\{ f_{-j} + f_{+j} \right\} \right]. \qquad (7.57)$$

For $n = 1$ to $n = 12$ coefficients a_j^n are computed and tabulated in the accompanying Fortran code using the slightly different notation as described in the unpublished report [11] included with the code. An excerpt is listed in Table 7.2 for $n = 1, \ldots, 6$. It is important to note that $2n + 1$ must not exceed the number of significant figures available on the computational resource, otherwise round-off error is amplified. The Fortran code calculates the coefficients to more than 20 significant figures.

FORTRAN CODE STIR

7.4 Gaussian Quadrature

7.4.1 Gauss-Legendre Formulas

The success of Newton-Cotes-type quadrature, as described in Sect. 7.3, depends on continued subdivision of the interval $[a, b]$ and application of a low-order polynomial approximation of the function to be integrated on each sub-interval. Gauss-type quadrature formulas use a global polynomial approximation to the function on the whole interval $[a, b]$. A Gaussian quadrature replaces the integral by a finite sum that gives the exact value of the integral subject to the requirement that the behavior

7.4 Gaussian Quadrature

Table 7.2 Coefficients

n	j	a_j^n	n	j	a_j^n
1	0	–	5	0	–
	1	+0.5000000000000000		1	+0.8333333333333333
2	0	–		2	−0.2380952380952381
	1	+0.6666666666666666		3	+0.5952380952380952(−1)
	2	−0.8333333333333333(−1)		4	−0.9920634920634921(−2)
				5	+0.7936507936507937(−3)
3	0	–	6	0	–
	1	+0.7500000000000000		1	+0.8571428571428571
	2	−0.1500000000000000		2	−0.2678571428571429
	3	+0.1666666666666667(−1)		3	+0.7936507936507937(−1)
4	0	–		4	−0.1785714285714286(−1)
	1	+0.8000000000000000		5	+0.2597402597402597(−2)
	2	−0.2000000000000000		6	−0.1803751803751804(−3)
	3	+0.3809523809523810(−1)			
	4	−0.3571428571428571(−2)			

of the function on $[a, b]$ is no more complicated than that of a polynomial of finite rank.

In view of the interval mappings given in (2.14), it suffices to discuss the normalized interval $[-1, +1]$, because the change of variable gives

$$\Im(f) = \int_a^b f(x)dx,$$

$$= \frac{1}{2}(b-a) \int_{-1}^{+1} f\left[\frac{1}{2}(b+a) + \frac{1}{2}(b-a)t\right] dt. \quad (7.58)$$

Therefore, in the remainder of this section the variable x is used and t is introduced only if the interval differs from $[-1, +1]$.

Consider the quadrature sum of (7.4) that is now viewed as a functional of $2n + 1$ parameters consisting of n weights $\{w_j\}_1^n$, n abscissas or nodes $\{x_j\}_1^n$, and the value of n. These parameters are adjusted so that the quadrature sum reproduces the integral exactly when the integrand is a polynomial of specific rank. For example, for a polynomial $p(x)$ of order ≤ 1, solve for the simplest possible quadrature sum that has only one weight w_1, and one abscissa x_1, such that the computed value of the integral gives the exact result. The required Gauss quadrature is

$$\int_{-1}^{+1} p(x)dx = w_1 p(x_1), \tag{7.59}$$

and to solve for two unknowns two linear equations are needed. The first equation is obtained by substituting the lowest order polynomial, $p(x) = 1$, into (7.59) and the second by substituting $p(x) = x$. After evaluating the integrals on the left-hand side for the respective monomials, the resulting equations are

$$2 = w_1 \times 1,$$
$$0 = w_1 \times x_1. \tag{7.60}$$

The solution is $w_1 = 2$, $x_1 = 0$, and the resulting quadrature formula is

$$\int_{-1}^{+1} p(x)dx = 2p(0), \tag{7.61}$$

which is exact as long as $p(x)$ is a polynomial with order no greater than one. To confirm this, consider the case when $p(x) = x^2$ is a quadratic polynomial. The error of the quadrature (7.61) is

$$\int_{-1}^{+1} x^2 dx - 2p(0) = \frac{2}{3}, \tag{7.62}$$

which is not zero. However, the next higher degree Gauss quadrature has two terms and four unknowns that may be chosen to make it exact for polynomials of order ≤ 3

$$\int_{-1}^{+1} p(x)dx = w_1 p(x_1) + w_2 p(x_2). \tag{7.63}$$

Four equations for the unknowns follow on substitution of $\{x^k\}_0^3$ into (7.63). This gives the respective series of equations

$$2 = w_1 + w_2,$$
$$0 = w_1 x_1 + w_2 x_2,$$
$$\frac{2}{3} = w_1 x_1^2 + w_2 x_2^2,$$
$$0 = w_1 x_1^3 + w_2 x_2^3, \tag{7.64}$$

7.4 Gaussian Quadrature

with the solution

$$w_1 = 1,$$
$$w_2 = 1,$$
$$x_1 = -\sqrt{\frac{1}{3}},$$
$$x_2 = +\sqrt{\frac{1}{3}}. \quad (7.65)$$

Obviously, the Gauss quadrature (7.63) is exact for all polynomials of order up to three, but how does it perform for more complicated functions? To test the quadrature consider again the example of (7.38) where the interval mapping onto $t \in [-1, +1]$ is required

$$\int_0^\pi \sin(x)dx = \frac{\pi}{2} \int_{-1}^{+1} \sin\left(\frac{\pi}{2} + \frac{\pi}{2}t\right) dt,$$

$$\approx \frac{\pi}{2} \sin\left(\frac{\pi}{2} - \frac{\pi}{2} \times \sqrt{\frac{1}{3}}\right) + \frac{\pi}{2} \sin\left(\frac{\pi}{2} + \frac{\pi}{2} \times \sqrt{\frac{1}{3}}\right),$$

$$\approx 1.93581957, \quad (7.66)$$

with an error of

$$E_2^G = 0.06418043. \quad (7.67)$$

This result is 1.47 times smaller than Simpson's rule estimate of (7.44) and requires only two function evaluations compared to three for Simpson's rule. This demonstrates the power and simplicity of Gauss-type quadrature for which a general theorem is the following.

Theorem 7.1 *If $q(x)$ is a polynomial of degree n such that*

$$\int_{-1}^{+1} q(x)x^m dx = 0, \quad m = 0, 1, \ldots, n-1, \quad (7.68)$$

with roots

$$-1 < x_1 < x_2 < \cdots < x_{n-1} < x_n < +1, \quad (7.69)$$

then the quadrature formula

$$\int_{-1}^{+1} f(x)dx = \sum_{i=1}^n w_i f(x_i) \quad (7.70)$$

gives the exact value of the integral whenever $f(x)$ is a polynomial of degree no greater than $2n - 1$.

To demonstrate this remarkable result, choose $f(x)$ so that it is a polynomial of degree at most $2n - 1$, and write it in factored form as

$$f(x) = p(x)q(x) + r(x) \tag{7.71}$$

where the divisor $q(x)$ has degree $\leq n$ and the quotient $p(x)$ and remainder $r(x)$ have degree $\leq n - 1$. Then, substitution of the form (7.71) into the integral gives

$$\int_{-1}^{+1} f(x)dx = \int_{-1}^{+1} p(x)q(x)dx + \int_{-1}^{+1} r(x)dx, \tag{7.72}$$

and the first term on the right-hand side is zero because the highest power of x in $p(x)$ is x^{n-1}, and (7.68) applies. Also, because $r(x)$ has degree at most $n - 1$, its integral is of order $\leq n$, and is exactly represented by an interpolatory polynomial constructed through n points. Therefore, in (7.72)

$$\int_{-1}^{+1} r(x)dx = \sum_{i=1}^{n} w_i r(x_i), \tag{7.73}$$

where the right-hand side is a polynomial of order n. But from (7.71) it follows that $f(x_i) = r(x_i)$, because the x_i are the roots of $q(x)$, and hence the result of (7.70) follows.

The polynomials $q(x)$ referred to above are the Legendre polynomials and the x_i used in the quadrature formulas are their roots. So that, for degree n in $q(x)$, there are n roots (7.69) and n weights w_i, corresponding to each root. Therefore, integrals of functions, $f(x)$, that behave like a polynomial whose degree does not exceed $2n - 1$, are given exactly by the quadrature sum of a degree n Gauss formula. Even if $f(x)$ does not satisfy this requirement, a sequence of quadratures with increasing n will generate a sequence of converging approximants for the integral. The first few abscissas, or nodes and their corresponding weights, are given in Table 7.3. As an example of their application, the corresponding sequence of approximants for the integral (7.38) is shown in Table 7.4, where strong convergence to the exact result with increasing n is evident.

The method used in (7.63) to solve for the weights becomes numerically unstable when n is large and therefore the following scheme is used to compute weights and abscissas. First the roots of the Legendre polynomial are determined from a root finding scheme such as those discussed in Chap. 6. If x_{jn} are the roots of the Legendre polynomial $P_n(x)$, of order n, indexed such that

$$-1 < x_{1n} < x_{2n} < \cdots < x_{n-1n} < x_{nn} < +1, \tag{7.74}$$

7.4 Gaussian Quadrature

Table 7.3 Gauss-Legendre quadrature

n	$\pm x_i$	w_i
2	0.577350269189626	1.000000000000000
3	0.000000000000000	0.888888888888889
	0.774596669241483	0.555555555555556
4	0.339981043584856	0.652145154862546
	0.861136311594053	0.347854845137454
5	0.000000000000000	0.568888888888889
	0.538469310105683	0.478628670499366
	0.906179845938664	0.236926885056189
6	0.238619186083197	0.467913934572691
	0.661209386466265	0.360761573048139
	0.932469514203152	0.171324492379170

Table 7.4 Gauss-Legendre application

n	$\int_0^\pi sin(x)dx$
2	1.935819574651137
3	2.001388913607745
4	1.999984228457722
5	2.000000110284471
6	1.999999999477270

then, for n either even or odd, the roots are symmetrically distributed about the origin. The roots are then used to obtain values for the weights from either of the expressions

$$w_{jn} = \frac{2(1 - x_{jn}^2)}{[nP_{n-1}(x_{jn})]^2},$$

$$= \frac{2}{(1 - x_{jn}^2)[P_n'(x_{jn})]^2}, \quad (7.75)$$

and as a numerical check it is useful to compute the sum

$$W_n = \sum_{j=1}^n w_{jn}, \quad (7.76)$$

which has the exact value of 2.

EXAMPLES 7.13–7.19
FORTRAN CODE PLZEROES, GAUSSLEG

7.4.2 Chebyshev Formulas

The Gauss-Legendre quadrature of (7.70) is so called because the weight function in the integrand is that for Legendre polynomials discussed in Sect. 2.4. A quadrature formula also exists for integrals with the Chebyshev weight function and what is special about the Chebyshev quadrature formula is that all the weights are the same for any order quadrature. There are two possible forms for such a quadrature formula. The first is the Gauss-Chebyshev form (see p. 115 of [7] where an expression for the remainder is also given)

$$\int_{-1}^{+1} f(x) \frac{dx}{\sqrt{1-x^2}} \approx \frac{\pi}{n} \sum_{j=1}^{n} f(x_j), \tag{7.77}$$

where the nodes are the roots (2.38) of the Chebyshev polynomial of order n

$$x_j = \cos\left(\frac{2j-1}{n}\frac{\pi}{2}\right). \tag{7.78}$$

The second form of the Chebyshev quadrature is that proposed by Clenshaw and Curtis [12]

$$\int_{-1}^{+1} f(x) \frac{dx}{\sqrt{1-x^2}} \approx \frac{\pi}{n} \sum_{j=0}^{n \prime\prime} f(x_j), \tag{7.79}$$

where the nodes are

$$x_j = \cos\left(j\frac{\pi}{n}\right), \tag{7.80}$$

and the double prime in (7.79) indicates that the series includes a factor of $\frac{1}{2}$ in both the first and last terms of the sum. Note that the Chebyshev weight $\frac{\pi}{n}$ is the same at all nodes on the right-hand side in the quadrature summation of both quadrature forms. Furthermore, the summation index of the Gauss-Chebyshev quadrature of (7.77) begins with 1 while that of the Clenshaw-Curtis form begins at 0. The difference in the two forms is that in (7.77), if the order, n, is changed then a new and different set of function values needs to be computed, because the roots of the Chebyshev polynomials are unique and different (except for the one at the origin for n odd). Whereas, in the Clenshaw-Curtis form, only the additional function values need to be evaluated, when changing from, say, n to $2n$.

As a simple demonstration, consider the integral

$$\int_{-1}^{+1} x^2 \frac{dx}{\sqrt{1-x^2}} = \frac{\pi}{2} \tag{7.81}$$

7.5 Special Case of an Oscillating Weight Function

Table 7.5 Gauss quadrature types and their weight functions

$[a, b]$	$\varphi(x)$	Name
$[-1, +1]$	1	Legendre
$[-1, +1]$	$\frac{1}{\sqrt{1-x^2}}$	Chebyshev
$[0, +\infty]$	e^{-x}	Laguerre
$[-\infty, +\infty]$	e^{-x^2}	Hermite

that is approximated by the Gauss-Chebyshev quadrature as

$$\int_{-1}^{+1} x^2 \frac{dx}{\sqrt{1-x^2}} \approx \frac{\pi}{n} \sum_{j=1}^{n} x_j^2, \qquad (7.82)$$

and for an order $n = 2$ this shows contributions at two roots of $T_2(x)$

$$\int_{-1}^{+1} x^2 \frac{dx}{\sqrt{1-x^2}} \approx \frac{\pi}{2} \left[\cos^2\left(\frac{1}{2}\frac{\pi}{2}\right) + \cos^2\left(\frac{3}{2}\frac{\pi}{2}\right) \right], \qquad (7.83)$$

while in the Clenshaw-Curtis form (7.79) the quadrature is

$$\int_{-1}^{+1} x^2 \frac{dx}{\sqrt{1-x^2}} \approx \frac{\pi}{2} \left[\frac{1}{2}\cos^2\left(0\frac{\pi}{2}\right) + \cos^2\left(1\frac{\pi}{2}\right) + \frac{1}{2}\cos^2\left(2\frac{\pi}{2}\right) \right], \qquad (7.84)$$

so that both quadrature forms give the exact result (7.81). This is not surprising because an n-point Gauss-type quadrature formula is exact if the function in the integrand is a polynomial of degree $\leq 2n - 1$.

Gauss-type quadrature formulas exist for each of the classical orthogonal polynomials and correspond to the respective intervals of orthogonality. Nodes and weights for these are to be found in references [13, 14] and tables [10]. A few of the simplest examples are identified by their names in Table 7.5, showing the interval of orthogonality $[a, b]$ and the respective weight function $\varphi(x)$, for integrals such as $\int_a^b f(x)\varphi(x)dx$. For details of the last two entries, see Sects. 7.4 and 7.5 of Krylov [7] while for recent studies of accuracy of quadrature formulas see [15, 16].

EXAMPLE 7.20

7.5 Special Case of an Oscillating Weight Function

The previous section discussed Gauss-type quadrature for integrals with integrands having a weight function $\varphi(x)$ that is always positive on the integration interval, as in the cases listed in Table 7.5. This section introduces a special case when the weight function is oscillatory.

A research project on nuclear scattering required quadrature for an integrand where the weight function $\varphi(x) = P_K(x)$, the Legendre polynomial of order K, defined in Sect. 2.4. These occur in the Legendre series of (2.51), where they are seen in the definition of the series expansion coefficients (2.52) and therefore may be considered to be the weight function in the integrand. This section summarizes some key results from [17] sufficient to describe the accompanying Fortran code that calculates the quadrature coefficients.

The inspiration for this work came from reading [7] and learning how to construct an interpolatory polynomial that turns out to be the Legendre polynomial. An additional advantage came from noting that a Clebsch-Gordan coefficient [18] occurs in the expression for the weights. Standard Fortran subroutines are available for their computation [19–21].

If the Legendre polynomial $P_K(x)$ has K roots on the interval $x \in [-1, +1]$ such that,

$$-1 < a_1 < a_2 < \cdots < a_{K-1} < a_K < +1, \qquad (7.85)$$

then the quadrature formula derived in [17] for the Legendre series coefficients of (2.52) is as follows:

$$g_K = \int_{-1}^{+1} f(x) P_K(x) dx,$$

$$= \sum_{k=1}^{n} B_k \left[f(x_k) - \sum_{j=1}^{K} A_{k,j} f(a_j) \right], \qquad (7.86)$$

where $\{x_k\}_1^n$ are the roots of the Legendre polynomial, $P_n(x), n \neq K$. The coefficients in (7.86) are defined as

$$B_k = \frac{2(1 - x_k^2)}{P_K(x_k)[n P_{n-1}(x_k)]^2} \sum_{m=0(2)}^{n-1 \leq 2K} P_m(x_k)[C(KKm; 000)]^2, \qquad (7.87)$$

$$A_{k,j} = \frac{P_K(x_k)}{(x_k - a_j)} \frac{(1 - a_j^2)}{K P_{K-1}(a_j)}, \qquad (7.88)$$

where the $C(KKm; 000)$ is the Clebsch-Gordan coefficient defined in [18]. From the property $P_s(-x) = (-1)^s P_s(x)$, it follows from (7.87), (7.88) that

$$B_{n-k+1} = (-1)^K B_k, \qquad (7.89)$$
$$A_{n-k+1, K-j+1} = A_{k,j}. \qquad (7.90)$$

Both sets of coefficients need to be evaluated once only and there is one proviso regarding the use of (7.87). When K is odd and $n = 2K + 1$, then $P_K(x)$ and $P_n(x)$ have a root in common at $x = 0$. However, calculation showed that the summation of (7.87) always gave a value of the order of the limit of accuracy whenever this case arises and therefore the coefficient was assumed to be zero.

The algebraic degree of precision of the interpolatory quadrature formula (7.86) is $2n+K-1$. In other words, the quadrature formula is exact when $f(x)$ is a polynomial of degree $\leq 2n+K-1$. The maximum degree of precision of this formula is attained when n takes its maximum allowable value, namely, $n = 2K+1$.

The computational form derived in [17] required evaluation of the the Clebsch-Gordan coefficient in (7.87). This is possible either by calls to the subroutines [19–21] or by applying a recurrence to the explicit expression for $[C(KKm; 000)]^2$ given by Rose (3.32) in [18], where $m = 2s$ to give

$$[C(KK2s; 000)]^2 = \frac{4s+1}{[2(K+s)+1]} \frac{[2(K-s)]!}{[2(K+s)]!} \left[\frac{(K+s)!(2s)!}{(K-s)!(s!)^2}\right]^2.$$

The recurrence relation in s follows from this expression upon taking the ratio of $[C(KK2s+2; 000)]^2$ to $[C(KK2s; 000)]^2$.

The roots $\{x_k\}_1^n$ of the Legendre polynomial, $P_n(x)$, are found by applying Newton's method (see Sect. 6.4) using the bounds derived in [22]

$$\cos\left[\frac{j_{0,n-k+1}}{[(n+1/2)^2+(c/4)]}\right] > x_k > \cos\left[\frac{j_{0,n-k+1}}{n+1/2}\right], \quad k=1,2,\ldots,n, \quad (7.91)$$

where $c = 1-(2/\pi)^2$, and $j_{0,i}$ is the ith zero of the Bessel function $J_0(x)$ (see Sect. 2.6) and are obtainable either from tables or a root finding method (see Chap. 6).

In the accompanying Fortran code, Newton's method is terminated when the correction is less than 10^{-24} and coefficients of the quadrature are computed from (7.87), (7.88) and tabulated. This quadrature formula was applied in a test case to compute the Legendre series coefficients of (4.102) for $z = 1$. The exact value is computed from the series in (4.104) that is truncated when the last term has a magnitude of 10^{-24} relative to the first term. Less than 15 terms is required to achieve this for $1 \leq K \leq 18$. Table 7.6 is a truncated summary of the higher precision results from the computation. The algebraic degree of precision in the quadrature formula is $5K+1$ and this accounts for the sharp improvement of the quadrature over the values of $1 \leq K \leq 9$. But this deteriorates as K approaches 18, due to increasing losses in figures of precision in the subtractions of (7.86).

FORTRAN CODE PLZEROES, GAUSSLEG

7.6 Exercises

1. To obtain three significant figures of precision in the integral $\int_0^\pi e^x sin(x)dx$ how many function evaluations are required for the trapezoid and Simpson's rules, respectively?

Table 7.6 Case study results

K	Exact I_K^P	Quadrature g_K	Error $\lvert I_K^P - g_K \rvert$
1	0.73575888234288	0.73536214416085	0.39(-03)
2	0.14312574025894	0.14312562825294	0.11(-06)
3	0.20130181048139(-01)	0.20130181036288(-01)	0.11(-10)
4	0.22144729219709(-02)	0.22144729219703(-02)	0.59(-15)
5	0.19992475040136(-03)	0.19992475040136(-03)	0.16(-19)
6	0.15300667555911(-04)	0.15300667555911(-04)	0.18(-23)

2. Compute the integral of the previous Exercise using Gauss-Legendre quadrature of orders $n = 2$ and $n = 3$. Compare with the results of Exercise 1 and comment on what you observe.
3. Determine the order of Gauss-Chebyshev quadrature required to obtain three significant figures of precision for the integral $\frac{2}{\pi} \int_{-1}^{+1} x \sin(m\pi x) (1-x^2)^{-1/2} dx$ when $m = 1, 2, 4$.
4. The integral (4.107) has the exact value (4.109). For $k = 1$ and $z = 1$ compute the numerical value of the integral with

    (a) the trapezoid rule using 14 evenly spaced function values;
    (b) Simpson's rule using 5 evenly spaced function values;
    (c) Gauss-Legendre quadrature of order $n = 2$ and $n = 3$; and
    (d) in each case show the exact value, quadrature value, absolute error, and relative error and comment on what you observe.

5. Repeat Exercise 4 with Gauss-Chebyshev quadrature of orders 3, 4 and 5 for $k = 1$ and $z = 1$.
6. For the integral $\int_{-1}^{+1} e^x \sin(x) dx$

    (a) apply both trapezoid and Simpson's quadrature using seven function values;
    (b) apply Gauss-Legendre quadrature for orders $n = 3$ and 5; and
    (c) comment on what you observe when comparing the results.

7. For the integral $\int_{-1}^{+1} e^{-4x^2} dx$

    (a) apply both trapezoid and Simpson's quadrature using nine function values;
    (b) apply Gauss-Legendre quadrature for orders $n = 3$ and 5; and
    (c) comment on what you observe when comparing the results.

7.7 Programming Problems

1. The Legendre series approximation (2.51) when $f(x) = sin(m\pi x)$ has only the odd-order Legendre polynomials appearing, because $f(x)$ in this case is an odd function. For $m = 1, 2$ and $k = 1, 3, \ldots, 11$ write a code that computes the coefficients of (2.52) using Trapezoid and Gauss-Legendre quadrature and attempt the maximum accuracy possible on the hardware and software platform available to you.

    (a) For Gauss-Legendre quadrature use the weights and abscissas in Table 7.3.
    (b) Compare the accuracy and cost of the two quadrature methods for the same k value by counting the number of function values required for a similar number of significant figures of precision.
    (c) Use the series expansion to evaluate the value at $x = 1$ and compare this to the exact value of $sin(m\pi)$.
    (d) For $m = 1, 2$ compute the L_p^2 error of (4.94) for the number of coefficients you deem sufficient for accuracy.

2. The spherical Bessel function $j_\ell(r)$ of order ℓ on the interval $r \in [0, a]$ has a Legendre series expansion for $r = ax$, $x \in [0, 1]$ of the form

$$j_\ell(ax) = \frac{\sqrt{\pi}}{2} \left(\frac{ax}{2}\right)^\ell \sum_{k=0}^{\infty} \left(k + \frac{1}{2}\right) g_k^\ell(a) P_k(x).$$

When $\ell = 0$ and $a = 1$ the function has the explicit expression $j_0(x) = \frac{sin(x)}{x}$ of (2.7), in which case the series has even-order polynomials only as is obvious from (2.23). Write a code to perform the following:

    (a) Apply Gauss-Legendre quadrature to compute values of the coefficients g_k^0 for the terms $g_k^0 P_{2k}(x)$ and $k = 0, .., 3$, in the definition of (2.52).
    (b) Compare the accuracy with the exact values of EXAMPLE 2.36 where $h_k = (k + \frac{1}{2}) g_k$ from the orthogonality of the Legendre polynomials in (2.47).

References

1. Gradshteyn IS, Ryzhik IM (1965) Tables of integrals series, and products, 4th edn. Academic Press, New York, NY
2. Klerer M, Grossman F (1971) A new table of indefinite integrals. Dover Publications, New York, NY
3. Wheelon AD (1968) Tables of summable series and integrals involving bessel functions. Holden-Day, San Francisco, CA

4. Piessens R et al (1983) QUADPACK a subroutine package for automatic integration. Springer-Verlag, Berlin
5. Davis PJ, Rabinowitz P (1967) Numerical integration. Blaisdell Publishing Company, Waltham, MA
6. Krommer AR, Ueberhuber CW (1998) Computational integration. Society for Industrial and Applied Mathematics, Philadelphia, PA
7. Krylov VI (1962) Approximate calculation of integrals. The Macmillan Company, New York, NY
8. Stroud AH (1971) Approximate calculation of multiple integrals. Prentice-Hall Inc, Englewood Cliffs, NJ
9. Stroud AH (1974) Numerical quadrature and solution of ordinary differential equations. Springer-Verlag, New York, NY
10. Abramowitz M, Stegun IA (eds) (1970) Handbook of mathematical functions. Dover Publications, New York
11. Delic G (1973) Formulae for numerical differentiation and integration. IKDA 73/8, Institut für Kernphysik, Technische Hochschule Darmstadt
12. Clenshaw CW, Curtis AR (1960) A method for numerical integration on an automatic computer. Numer Math 2:197–205
13. Olver FWJ, Lozier DW, Boisvert RF, Clark CW (eds) (2010) NIST handbook of mathematical functions. NIST and Cambridge University Press, UK
14. Spanier J, Oldham KB (1987) An atlas of functions. Hemispere Publishing Corporation, New York
15. Fornberg B (2021) Improving the accuracy of the trapezaoid rule. SIAM Rev 63:167–180
16. Trefethen LN (2022) Exactness of quadrature formulas. SIAM Rev 64:132–150
17. Delic G (1974) The legendre series and a quadrature formula for its coefficients. J Comput Phys 14:254–268
18. Rose ME (1957) Elementary theory of angular momentum. John Wiley and Sons, New York
19. Tamura T (1970) Angular momentum coupling coefficients. Comput Phys Commun 1:337–342
20. Tamura T (1971) Erratum. Comput Phys Commun 2:174
21. Wills JG (1971) On the evaluation of angular momentum coupling coefficients. Comput Phys Commun 2:381–2
22. Szegö G (1936) Inequalities for the zeros of legendre polynomials and related functions. Trans Am Math Soc 39:1–17

Two-Dimensional Numerical Integration

8.1 Definition of the Integral

In this section, the definition of the definite integral is extended to a two-dimensional domain and is defined as

$$\int_a^b \int_c^d f(x, y)\, dx dy = F(a, b, c, d), \tag{8.1}$$

where $F(x, y)$ is a function whose derivative is $f(x, y)$, defined on a rectangle $\mathbb{R}$ spanned by $x \in [a, b]$, $y \in [c, d]$, where it is continuous and bounded. The Riemann integral may be defined on a grid in $\mathbb{R}$ as described in Sect. 1.7 of [1]. The resulting number $F(a, b, c, d)$ may be thought of as the volume under the surface $f(x, y) \subset \mathbb{R}$. In general, the domain need not be rectangular and may be mapped into a unit square (if a suitable mapping exists). Therefore, attention here is limited to the rectangular case only.

In the following sections a few examples show how the definite integral is evaluated approximately by a numerical technique that applies results from the previous chapter. This uses products of quadrature formulas in each of the two dimensions to form a sum similar to that in (7.4). An alternative is to apply minimal point formulas to economize on the number of function evaluations. The function values $f(x, y)$ are computed at n discrete pairs of points and respectively multiplied by each of the n discrete weights w_i. The discussion in the next section is limited to product formulas of Simpson's rule and Gauss-Legendre quadrature. Refinement of the numerical

---

Supplementary Information The online version contains supplementary material available at https://doi.org/10.1007/978-3-031-90178-2_8

estimate is achieved by partitioning the domain $\mathbb{R}$ into smaller rectangles and applying the same product formula to each. This product formula approach can become expensive by increasing the number of function evaluations required, especially for integrals in higher dimensional spaces (see Chap. 5 of [1]). For this reason the subsequent section introduces minimal point formulas in the two-dimensional case and the last section discusses a statistical approach for higher dimensional integrals.

8.2 Product Formulas

8.2.1 Simpson's Rule

Following Davis and Rabinowitz (p. 131 of [1]) the product of two Simpson rules for the rectangular domain $\mathbb{R}$ has $3 \times 3 = 9$ points to integrate exactly all linear combinations of 16 monomials $x^i y^j$ for $0 \leq i, j \leq 3$, written as follows in the notation of (7.37) extended to two dimensions

$$\begin{aligned} S^{\mathbb{R}} &= f_{1,1} + f_{1,3} + f_{3,1} + f_{3,3} \\ &+ 4 \left\{ f_{1,2} + f_{2,1} + f_{3,2} + f_{2,3} \right\} \\ &+ 16 f_{2,2}, \end{aligned} \qquad (8.2)$$

and

$$\int_a^b \int_c^d f(x,y)\, dx\, dy = \frac{h_x}{3} \frac{h_y}{3} \times S^{\mathbb{R}}, \qquad (8.3)$$

where, following the notation of (7.37), $h_x = (b-a)/2$, and $h_y = (d-c)/2$, with

$$f_{i,j} = f(x_i, y_j), \qquad (8.4)$$

and

$$\begin{aligned} x_1 &= a, \\ x_2 &= a + h_x, \\ x_3 &= b, \\ y_1 &= c, \\ y_2 &= c + h_y, \\ y_3 &= d. \end{aligned} \qquad (8.5)$$

The first estimate applies the product of Simpson's rule to the entire rectangle $\mathbb{R}$, and the next refinement subdivides it into four tiles of equal area.

EXAMPLES 8.1–8.4

8.2.2 Gauss-Legendre Quadrature

A product formula for (8.1) of the Gauss-Legendre quadrature type in two dimensions (2-D), on the unit square, analogous to (7.70), takes the form

$$\int_{-1}^{+1}\int_{-1}^{+1} f(x,y)dxdy = \sum_{i=1}^{n}\sum_{j=1}^{m} w_i w_j f(x_i, y_j), \quad (8.6)$$

where the indices n, m represent the order of the quadrature to give the exact value of the integral, whenever $f(x, y)$ is a polynomial of degree at most $2n - 1, 2m - 1$ in the respective dimensions. The lowest order Gauss-Legendre quadrature has the nodes and weights defined in (7.65) and higher orders are tabulated in Table 7.3 and evaluated by the Fortran program GAUSSLEG (see Sect. 7.4.1). The EXAMPLES demonstrate applications and how a mapping into the unit square is applied when needed. The efficiency of any quadrature is measured in terms of the number of function evaluations for a specified precision. Therefore note should be made that, with this criterion, Gauss-Legendre quadrature is superior to fixed-step formulas, such as Simpson's rule.

EXAMPLES 8.5–8.8

8.3 Minimal Point Formulas

Product quadrature formulas for integrals in multiple dimensions are useful but can become expensive in terms of the number of function evaluations they require. The author's search for alternative solutions came while working on code for nuclear scattering calculations. Initially, these applied product formulas of Newton-Cotes-type with fixed-step one-dimensional (1-D) quadrature rules [2]. However, in later work [3], perfectly symmetric minimal point 2-D formulas were applied. Such alternatives are based on polynomial interpolation familiar in 1-D cases and extended to multiple integrals. A comprehensive discussion is given in the publications by Stroud [4–7]. Stroud tabulates numerous formulas (Chap. 8 of [6]) for the 18 regions defined in Chap. 6, as well as the theory for producing them in Chap. 2. In this section the focus is on perfectly symmetric minimal point 2-D formulas previously applied by the author [3], as these are more than sufficient for the square region to which they are applied.

Minimal point, fully symmetric formulas, have a quadrature of the form

$$\int_{-1}^{+1}\int_{-1}^{+1} f(x,y)dxdy = \sum_{i=1}^{n} w_i f(x_i, y_i), \quad (8.7)$$

Table 8.1 Tyler's 2-D quadrature

x	y	w
0.92582009	0.0	0.24197531
0.38055443	0.38055443	0.52059292
0.80597978	0.80597978	0.23743177

where the nodes $\{x_i, y_i\}_1^n$ form a discrete symmetric set on the unit square $x \in [-1, +1]$, $y \in [-1, +1]$. Specifically, a 12-point, degree 7, formula (pp. 252–253 in [6]) is due to Tyler (as shown in Fig. 4 of [8]) and is reproduced in Table 5 of [9]. Also, applied was a 28-point, degree 11, formula (pp. 259–260 in [6]) and 48-point, degree 15, formula (p. 261 in [6]) from Table 1 of Rabinowitz and Richter [10]. The reason for this choice was to form a hierarchy of increasing precision from squares with a ratio of side length 1:2:4 in [3]. These were used to pack a parallelogram shaped region along a 2-D diagonal with the 48-point quadrature applied to the largest squares on the diagonal and the other two successively smaller squares populating the interstices outside the largest squares. A better approach would have been to map the parallelogram into a rectangle, but neither packing method is described here. Instead, as a demonstration, the simplest example shows how symmetry is applied in the 12-point formula to (8.7). The 12-point formula is shown in Table 8.1 where the generators for 2-D coordinates are listed in the first two columns and corresponding weights in the third.

To show how full symmetry is applied to the values in Table 8.1, the 2-D quadrature sum in (8.7) has $n = 12$ terms as follows:

$$\int_{-1}^{+1}\int_{-1}^{+1} f(x,y)\,dx\,dy = \sum_{i=1}^{3} w_i \left[f(x_i, y_i) + f(-x_i, y_i) + f(x_i, -y_i) + f(-x_i, -y_i) \right]$$

(8.8)

with similar application to the 28 and 48 point formulas as is shown in the EXAMPLES for 2-D, 4-D and 6-D cases. In the last two cases, EXAMPLE 8.16 shows how the Compressed Column (CC) sparse matrix storage scheme of Sect. 5.4 reduces the floating point operation count from 5184 to 2304 (4-D) and from 373248 to 110592 (6-D), respectively. This provides an important improvement in efficiency especially if the algorithm is applied repeatedly.

EXAMPLES 8.9–8.16

8.4 The Monte Carlo Method

When the multiplicity of the integrals increases beyond 2-D, or when the volume spanned is irregular, then the above described formulas can become expensive, or difficult to apply. In such cases a statistical method known as the Monte Carlo Method may be applied. Numerical applications of this method were first introduced in [11]

8.4 The Monte Carlo Method

and some expositions are found in Chap. 6 of Stroud [6], the short monograph of Sobol' [12] and Sect. 6 of Haber [13]. Stroud also includes useful Fortran code.

To describe this statistical method consider the multiple integral defined as

$$\Im(f) \equiv \int \cdots \mathbb{R}_{\mathbb{N}} \int f(x_1, \ldots x_N) \, dx_1 \cdots dx_N, \qquad (8.9)$$

where, for each N, a set of points $\{v_i\}_1^N$ is randomly chosen and uniformly distributed in $\mathbb{R}_{\mathbb{N}}$. These are viewed as independent discrete random variables with probabilities $\{f(v_i)\}_1^N$ associated with each. These probabilities are assumed to be distributed uniformly and are all positive and non-zero when they are inside the region $\mathbb{R}_{\mathbb{N}}$ and zero outside. The probability of them being found in any particular sub-region of $\mathbb{R}_{\mathbb{N}}$ with volume V is directly proportional to the volume of that sub-region [13]. In this case, for large N, $\Im(f)$ is assumed to have the mean value

$$\frac{V}{N} \sum_{i=1}^{N} f(v_i), \qquad (8.10)$$

with variance

$$\gamma^2 = \frac{1}{N} \left[\sum_{i=1}^{N} \{f(v_i)\}^2 - \frac{1}{N} \left(\sum_{i=1}^{N} f(v_i) \right)^2 \right], \qquad (8.11)$$

and a standard deviation $\gamma/\sqrt{N}$. The classical Monte Carlo method for estimating a multiple integral is introduced as

$$\Im(f) \approx Q_{0,N}, \qquad (8.12)$$

with

$$Q_{0,N}(f) \equiv \frac{V}{N} \sum_{i=1}^{N} f(v_i), \qquad (8.13)$$

where V is the N-dimensional volume of $\mathbb{R}_{\mathbb{N}}$. Hence, $Q_{0,N}$ in (8.13) is a random variable with mean (8.10) and variance (8.11). Because the standard deviation does not decrease rapidly with increasing N, methods have been developed to decrease the error in the approximation (8.10). For a review of such methods see Haber [13], but for the purposes of this short introduction the standard deviation is used as an error estimate of the approximation (8.13). For further discussion of error estimates, see Sect. 5.9 of [10] and Sect. 5.5 of [14] that also has some Fortran code samples. The EXAMPLES are 4-D and 6-D cases with the number of samples ranging from 10^1 to 10^6.

EXAMPLES 8.17 and 8.18

8.5 Exercises

1. For the integral $\int_0^\pi \int_0^\pi e^{x+y} \sin(x)\sin(y)dxdy$, obtain three significant figures of precision using product quadrature formulas of trapezoid and Simpson's rules, respectively. How many function evaluations are required for each method?
2. Compute the integral of the previous Exercise using product Gauss-Legendre quadrature of orders $n = 2$ and $n = 3$. Compare with the results of Exercise 1 and comment on what you observe.
3. For the integral $\int_0^\pi \int_0^\pi e^{x+y} \sin(x)\sin(y)dxdy$

    (a) apply the minimal 12-point Tyler formula of Table 8.1,
    (b) apply the Monte Carlo statistical method with 12 function evaluations (if necessary subdivide the area under the 2-D region into smaller tiles),
    (c) comment on what you observe when comparing the results,
    (d) compare the results of (a) and (b) with those of Exercises 1 and 2 for a similar number of function evaluations.

8.6 Programming Problems

1. The product of spherical Bessel functions $j_\ell(r)j_\ell(s)$ of order ℓ on the 2-D region $R \in [0, a] \times [0, a]$ has a Legendre series expansion for $r = ax$, $s = ay$, $x, y \in [0, 1]$ in the form

$$j_\ell(ax)j_\ell(ay) = \frac{\pi}{2^2}\left(\frac{ax}{2}\right)^\ell \left(\frac{ay}{2}\right)^\ell$$
$$\times \sum_{k=0}^\infty \sum_{m=0}^\infty \left(k+\frac{1}{2}\right)\left(m+\frac{1}{2}\right) g_k^\ell(a) e_m^\ell(a) P_k(x) P_m(y)$$

and for $\ell = 0$, and $a = 1$ it has the explicit expression $j_0(x)j_0(y) = \frac{sin(x)}{x}\frac{sin(y)}{y}$ of (2.7), in which case both series have even-order polynomials only as is obvious from (2.23). Write a code to perform the following:

(a) Apply Gauss-Legendre quadrature to compute values of the coefficients g_k^0, e_m^0 for the terms $g_k^0 P_{2k}(x)$ and $e_m^0 P_{2m}(x)$, $k, m = 0, ..., 3$, in the definition of (2.52).
(b) Compare the accuracy with the exact values of EXAMPLE 2.36 where $h_k = (k+\frac{1}{2})g_k$, and similarly for e_m^0 from the orthogonality of the Legendre polynomials in (2.47).
(c) Repeat (a) and (b) for the minimal 12-point Tyler formula of Table 8.1.
(d) Repeat (a) and (b) for the minimal 28-point and 48-point formulas (see EXAMPLES 8.9, 8.12, 8.14).

References

1. Davis PJ, Rabinowitz P (1967) Numerical integration. Blaisdell Publishing Company, Waltham, MA
2. Delic G, Robson BA (1970) Dwba calculations for (d, p) reactions including the d-state of the deuteron. Nucl Phys A 156:97–104
3. Delic G, Robson BA (1974) The deuteron d-state and exact finite-range dwba calculations for 52cr(d, p)53cr. Nucl Phys A 232:493–501
4. Stroud AH (1961) A bibliography on approximate integration. Math Comput 15:52–80
5. Stroud AH (1969) Integration formulas and orthogonal polynomials for two variables. SIAM J Numer Anal 6:222–229
6. Stroud AH (1971) Approximate calculation of multiple integrals. Prentice-Hall Inc, Englewood Cliffs, NJ
7. Stroud AH (1974) Numerical quadrature and solution of ordinary differential equations. Springer-Verlag, New York, NY
8. Tyler GW (1953) Numerical integration of functions of several variables. Can J Math 5:393–412
9. Franke R (1971) Obtaining cubatures for rectangular and other planar regions by using orthogonal polynomials. Math Comput 25:803–817
10. Rabinowitz P, Richter N (1969) Perfectly symmetric two-dimensional integration formulas with minimal number of points. Math Comput 23:765–779
11. Metropolis IN, Ulam S (1949) The monte carlo method. J Am Stat Assoc 44:135–41
12. Sobol' IM (1974) The Monte Carlo method. The University of Chicago Press, Chicago, IL
13. Haber S (1970) Numerical evaluation of multiple integrals. SIAM Rev 12:481–526
14. Brandt S (1976) Statistical and computational methods in data analysis. North-Holland Publishing Company, Amsterdam

Numerical Solution of Ordinary Differential Equations

9.1 Background

The solution of Ordinary Differential Equations (ODEs) [1] and their analogues, Partial Differential Equations (PDEs) [2,3], have a long history. The basic types of equations maybe identified in their simplest forms as the Laplace equation, the wave equation, and the diffusion equation (or heat equation) [4] and the latter two are examined in this chapter. The history of solution methods has been strongly affected by the development of computers and numerical solution methods, but only some special cases and (numerical) solution methods are discussed. The number and type of numerical algorithms are varied. However, all attempt to project the equation of interest into a discrete form on a physical spatial (or temporal) grid that is then solvable as a finite rank matrix problem. This discretization implies that continuous functions and operators on them are smooth and well defined on the interval of interest.

The earliest discretization method of the computer era was the finite difference method [5,6] that uses finite difference arithmetic (see Sect. 3.8) to represent operators in ODEs and polynomial interpolation of discrete data (see Chap. 3). Following this, the method of finite elements [7–11] and spline interpolation [12,13] introduced specific discretization methods using low-order polynomials on sub-intervals. Surveys and comparison of numerical solution methods also appeared [14–16] with studies of stability and convergence behavior [17]. Of particular interest in this chapter are Galerkin methods [18] which come in either local, low-order polynomial approximation, or global in the form of spectral functions defined on the whole interval of

---

Supplementary Information The online version contains supplementary material available at https://doi.org/10.1007/978-3-031-90178-2_9

approximation [19–22]. The spectral method discussed here is based on the author's prior work [23] using trigonometric or polynomial functions defined globally over the interval of interest.

9.2 Definitions

Returning to the general form of an operator equation (5.1), this chapter investigates the specific case when the left-hand side is a differential operator. Attention in this chapter is limited to two specific forms shown here in simplified examples.

This first form, in one spatial dimension, is the heat (or diffusion) PDE first derived by Fourier [24]

$$\frac{\partial u}{\partial t} = \kappa \frac{\partial^2 u}{\partial r^2}, \qquad (9.1)$$

for which he also demonstrated that a trigonometric series (see Sect. 2.5) could be applied to represent the solution $u(r, t)$ on a region $r \in [0, R]$, at any later time $t > 0$, with κ a constant in the linear case.

The second form is the wave equation derived by d'Alembert[1] [25] in his prolific studies of sound vibrations in strings, shown here as a one-dimensional PDE

$$\frac{\partial^2 u}{\partial t^2} = c^2 \frac{\partial^2 u}{\partial r^2}. \qquad (9.2)$$

In both cases, the coefficient on the right-hand side is a constant for the linear problem.

These two types of operator equations represent (in simplified form) the heat equation and wave equation. They are examples of evolution equations, or ordinary differential equations, where the solution propagates in time. In the former case, the solution decays (or grows) exponentially with time and, in the latter case, the solution will either be a stationary wave, or propagates as a wave envelope, depending on the boundary conditions on the interval.

Both types of operator equations shown above have a long history and, in many cases, analytical solutions are known [15, 26, 27], whereas numerical solution is necessary in other cases [8, 14, 16, 17, 28, 29]. Some studies propose parallel algorithms [30, 31], while others include Fortran code [7, 32, 33]. The following sections describe a few specific forms of the heat and wave equations with methods of solution studied by the author during his research.

9.3 Numerical Solution Methods on a Physical Grid

In the first instance consider, the following one-dimensional (1-D) form of a differential equation

$$\frac{du}{dx} = f(x, u), \qquad (9.3)$$

---

[1] French mathematician, encyclopedist and philosopher, 1717–1783.

9.3 Numerical Solution Methods on a Physical Grid

where the left-hand side is a derivative with respect to one spatial coordinate and the right-hand side is a function of $x, u(x), x \in [0, X]$. A general solution is

$$u(x) = \int f(x, u) dx + C, \quad (9.4)$$

where C is an unknown constant. To find a specific solution, an initial value of $u(x)$ is required for some $x = x_0$

$$u_0 = u(x_0). \quad (9.5)$$

The problem of solving (9.3) is known as an Initial Value problem (IVP) (see Chap. 14 of [34]), or Chap. 9 of [35]).

9.3.1 Euler's Method

One of the simplest numerical methods is that of Euler which combines the approximation of the derivative in the form of (2.31) with the Taylor series in (2.36) to yield the single step approximation with a step size h

$$\begin{aligned} u_{n+1} &= u_n + h u'_n, \\ &= u_n + h f(x_n, u_n), \end{aligned} \quad (9.6)$$

where the prime denotes the first derivative with respect to the argument x and the common notation used is

$$\begin{aligned} u_n &= u(x_n), \\ x_n &= x_0 + nh, \\ u_0 &= u(x_0). \end{aligned} \quad (9.7)$$

From (2.36) it is seen that Euler's method is stable and has an error $\mathcal{O}(h^2)$. Note that truncation of the series (2.36) after the h^2 term gives a second-order difference equation

$$u_{n+2} = 2u_{n+1} - u_n + h^2 u''_n, \quad (9.8)$$

with a truncation error $\mathcal{O}(h^3)$, but this choice may experience stability issues (see Sect. 6.7 and Chap. 14 of [34]).

EXAMPLES 9.1–9.3: EULER FORMULA

9.3.2 Multi-step Methods

Multi-step methods offer powerful tools for integration of IVPs for ODEs where a sequence of approximations $\{u_n\}_1^N$ to $\{u(x_n)\}_1^N$ are generated by a linear multi-step integration formula of the general type

$$u_{n+1} = \sum_{i=1}^{k} \alpha_i u_{n+1-i} + h \sum_{i=0}^{k} \beta_i u'_{n+1-i} + T(x, h). \qquad (9.9)$$

The $2k+1$ parameters $\{\alpha\}_1^k$, $\{\beta\}_0^k$ are constants independent of h, and $T(x,h)$ is the error term due to the truncation of the series. If $k=1$ then this is a single step method. If $\beta_0 = 0$, the formula is explicit and if $\beta_0 \neq 0$, it is implicit and requires a value for

$$u'_{n+1} = \left(\frac{du}{dx}\right)_{x=x_{n+1}}. \qquad (9.10)$$

Two classes of methods for obtaining the integration formulas (9.9) are the Runge-Kutta (RK) and predictor-corrector (PC). The RK method generates u', $f(x, u)$ at the points x_n and also at values of x, u in between successive x_n, u_n, whereas the PC method generates u', $f(x, u)$ only at the points x_n. Each method has associated with it an error, $T(x, h)$, $x \in [x_{n-k}, x_{n+1}]$, that is, the error produced at step $n+1$ if u_j and u'_j on the right-hand side of (9.9) are known without error.

Sources of error are truncation error and round-off error. The former occurs because the right-hand side is not an exact representation of u_{n+1} due to the use of a finite interpolating polynomial for u. Round-off error is due to finite precision arithmetic on the computer resource (see Chap. 1). The calculation of u_{n+1} from inexact values of u_n, u'_n propagates error through subsequent steps. The source of error is studied under the question of stability of a given integration formula. Detailed error and stability analyses for the integration formula used here are to be found in [32, 36–38].

9.3.3 Runge-Kutta

The RK method evaluates $f(x, u)$ within the interval (x_n, u_n) to (x_{n+1}, u_{n+1}) at intermediate points in a general single step formula of the type

$$u_{n+1} = u_n + \sum_{j=1}^{v} w_j r_j, \qquad (9.11)$$

where w_j are weight coefficients and v is the number of evaluations of $f(x, u)$ with

$$r_j = h f\left(x_n + c_j h, u_n + \sum_{m=1}^{j-1} a_{jm} r_m\right), j = 1, 2, \cdots, v, \qquad (9.12)$$

where, by definition, $c_1 = 0$ and $r_1 = h f(x_n, u_n)$. Different sets of parameters $\{w_j\}$, $\{c_j\}$, $\{a_{jm}\}$ determine different RK integration formulas and each specifies (x, u) values at which $f(x, u)$ is evaluated. Table 9.1 shows these parameters represented in the form known as the Butcher Table [39] in honor of Butcher [36],

9.3 Numerical Solution Methods on a Physical Grid

Table 9.1 Butcher table

c_j	(p, v)	B			
0					
c_2	a_{21}				
c_3	a_{31}	a_{32}			
$\vdots$	$\vdots$	$\vdots$			
c_v	a_{v1}	a_{v2}	$\cdots$	a_{vv-1}	
	w_1	w_2	$\cdots$	w_{v-1}	w_v

where p is the order of the formula and v is the number of terms in (9.11). In the case of an explicit RK method (the only cases considered here), the table is lower triangular with the first column as the vector of coefficients c and the bottom row is the transpose of the vector of weights w^T. The explicit form requires only terms with r_m, $m = 1, \cdots, j - 1$ in (9.11) and the coefficients $\{c_j\}$ have the property $c_j = \sum_{m=1}^{j-1} a_{jm}$.

For specific examples of (9.11), (9.12) the second-order case is

$$\begin{aligned} r_1 &= hf(x_n, u_n), \\ r_2 &= hf(x_n + c_2 h, u_n + a_{21} r_1), \\ u_{n+1} &= u_n + w_1 r_1 + w_2 r_2, \\ c_2 &= a_{21}, \end{aligned} \tag{9.13}$$

and the third-order case is

$$\begin{aligned} r_1 &= hf(x_n, u_n), \\ r_2 &= hf(x_n + c_2 h, u_n + a_{21} r_1), \\ r_3 &= hf(x_n + c_3 h, u_n + a_{31} r_1 + a_{32} r_2), \\ u_{n+1} &= u_n + w_1 r_1 + w_2 r_2 + w_3 r_3, \\ c_2 &= a_{21}, \\ c_3 &= a_{31} + a_{32}, \end{aligned} \tag{9.14}$$

while the fourth-order case is

$$\begin{aligned} r_1 &= hf(x_n, u_n), \\ r_2 &= hf(x_n + c_2 h, u_n + a_{21} r_1), \\ r_3 &= hf(x_n + c_3 h, u_n + a_{31} r_1 + a_{32} r_2), \\ r_4 &= hf(x_n + c_4 h, u_n + a_{41} r_1 + a_{42} r_2 + a_{43} r_3), \\ u_{n+1} &= u_n + w_1 r_1 + w_2 r_2 + w_3 r_3 + w_4 r_4, \\ c_2 &= a_{21}, \end{aligned}$$

Table 9.2 Second order

c_j	(2,2)		$B = 2$
0			
2	2		
	1	1	

Table 9.3 Third order

c_j	(3,3)			$B = 24$
0				
16	16			
16	0	16		
	6	9	9	

$$c_3 = a_{31} + a_{32},$$
$$c_4 = a_{41} + a_{42} + a_{43}.$$

The parameters are tabulated in [40] for RK orders 2–6 and in the accompanying FORTRAN CODE RKPC. Examples of second- to fourth-order cases are shown in Tables 9.2, 9.3, and 9.4 where formulas are classified by the pair (p, v) with p the order of the formula and v the number of terms in (9.11). The values in the tables have been multiplied by a suitable scaling constant B so that they appear as integers. It should be noted that these RK solutions are not unique and alternative explicit forms may be found in sources such as Butcher [36] and Hairer et al. [32].

From Table 9.2 the second-order case of (9.13) appears as follows, on noting the division of c_2, w_1, w_2, by the normalization factor $B = 2$

$$r_1 = hf(x_n, u_n),$$
$$r_2 = hf(x_n + 1 \times h, u_n + 1 \times r_1), \qquad (9.15)$$
$$u_{n+1} = u_n + \frac{1}{2}r_1 + \frac{1}{2}r_2,$$
$$c_2 = 1.$$

The third-order case of (9.14), again noting the division of c_2, c_3, w_1, w_2, w_3 by the normalization factor $B = 24$ (Table 9.3), is

$$r_1 = hf(x_n, u_n),$$
$$r_2 = hf(x_n + \frac{2}{3}h, u_n + \frac{2}{3}r_1),$$
$$r_3 = hf(x_n + \frac{2}{3}h, u_n + 0 \times r_1 + \frac{2}{3}r_2), \qquad (9.16)$$
$$u_{n+1} = u_n + \frac{1}{4}r_1 + \frac{3}{8}r_2 + \frac{3}{8}r_3,$$

9.3 Numerical Solution Methods on a Physical Grid

Table 9.4 Fourth order

c_j	(4, 4)	$B = 24$		
0				
8	8			
16	−8	24		
24	24	−24	24	
	3	9	9	3

$$c_2 = \frac{2}{3},$$
$$c_3 = 0 + \frac{2}{3},$$

while the fourth-order case shown in Table 9.4 is

$$\begin{aligned}
r_1 &= hf(x_n, u_n), \\
r_2 &= hf(x_n + \frac{1}{3}h, u_n + \frac{1}{3}r_1), \\
r_3 &= hf(x_n + \frac{2}{3}h, u_n - \frac{1}{3}r_1 + 1 \times r_2), \\
r_4 &= hf(x_n + 1 \times h, u_n + 1 \times r_1 - 1 \times r_2 + 1 \times r_3), \quad (9.17) \\
u_{n+1} &= u_n + \frac{1}{8}r_1 + \frac{3}{8}r_2 + \frac{3}{8}r_3 + \frac{1}{8}r_4, \\
c_2 &= \frac{1}{3}, \\
c_3 &= \frac{2}{3}, \\
c_4 &= 1.
\end{aligned}$$

Explicit RK formulas in each of the orders listed above are not unique and different families are known [41]. These tables and examples are taken from [38] and include larger orders from [40] and in the accompanying FORTRAN CODE RKPC, from that publication.

The EXAMPLES and the Fortran code examine results of RK formulas of order 2, 3, and 4, with the Euler formula (sometimes known as RK of order 1), for the following three simple systems from Lapidus and Seinfeld [38]

$$\begin{aligned}
u' &= -u \\
u' &= +u \quad (9.18) \\
u' &= 1 - u
\end{aligned}$$

with the respective initial values

$$u_0 = 1,$$
$$u_0 = 1, \qquad (9.19)$$
$$u_0 = 0.$$

The ODEs of (9.18) have the respective analytical solutions

$$u(x) = e^{-x},$$
$$u(x) = e^{+x}, \qquad (9.20)$$
$$u(x) = 1 - e^{-x},$$

and these are used to compare against the RK predicted values as instructive demonstrations of how the error is affected by the step size choice and successive increases in order.

EXAMPLES 9.4–9.6: RUNGE-KUTTA ORDER 2
EXAMPLES 9.7–9.9: RUNGE-KUTTA ORDER 3
EXAMPLES 9.10–9.12: RUNGE-KUTTA ORDER 4
FORTRAN CODE RKPC

9.3.4 Predictor-Corrector

In the PC method a predictor formula of the type in (9.9) is used to approximate the value $\bar{u}_{n+1}$ for u_{n+1}. Then a corrector formula of the same type uses this value to improve the estimate for u_{n+1}. The basic steps in the PC method are

1. Prediction of a value $\bar{u}_{n+1}$ for u_{n+1}.
2. Evaluation of the derivative from the ODE $\bar{u}'_{n+1} = f(x_{n+1}, \bar{u}_{n+1})$.
3. Correction of $\bar{u}_{n+1}$.
4. Evaluation of u'_{n+1}.

The method is referred to as the PE(CE)^S method in [38] because steps 3 and 4 are iterated S times. In an alternative method, PEC, step 4 is omitted but the function value using the predicted $\bar{u}_{n+1}$ is used in the next cycle. Steps 2 and 3 may be iterated to give a P(EC)^S method. Thus, with $S = 1$ the PEC method gives u_{n+1} and $\bar{u}'_{n+1}$, while the PECE method gives u_{n+1} and u'_{n+1}. Among features listed in [40] two are worth repeating here. The error estimate for the truncation error is easily obtained as

$$\epsilon_{n+1} = u_{n+1} - \bar{u}_{n+1}, \qquad (9.21)$$

and may be used as an adaptive PC algorithm where the step size is adjusted during the integration.

The PC formulas listed in [40] are of the Adams form with $\alpha_1 = 1, \alpha_i = 0, i > 1$, in (9.9) and the predictor is the Bashforth form with $\beta_0 = 0$ while the corrector form

9.3 Numerical Solution Methods on a Physical Grid

Table 9.5 Predictors

k	d_k	$d_k\beta_1$	$d_k\beta_2$	$d_k\beta_3$	$d_k\beta_4$
2	2	3	−1		
3	12	23	−16	5	
4	24	55	−59	37	−9

Table 9.6 Correctors

k	d_k	$d_k\beta_0$	$d_k\beta_1$	$d_k\beta_2$	$d_k\beta_3$	$d_k\beta_4$
1	2	1	1			
2	12	5	8	−1		
3	24	9	19	−5	1	
4	720	251	646	−264	106	−19

is the Moulton form with $\beta_0 \neq 0$. A subset of these is shown in Tables 9.5 (Adams-Bashforth predictors) and 9.6 (Adams-Moulton correctors), respectively, where d_k is a multiplier used to normalize the coefficient entries for display as integers. These tables and examples are taken from [38] and include larger orders from [40] in the accompanying FORTRAN CODE RKPC from that publication.

The PC method is applied to the simple examples of (9.18) for orders 3 and 4 with only one iteration. Again the error results need to be compared with the RK method for each of these examples. The RKPC code allows for multiple iterations through an input parameter.

EXAMPLES 9.13–9.15: PREDICTOR- CORRECTOR FORMULA ORDER 3
EXAMPLES 9.16–9.18: PREDICTOR- CORRECTOR FORMULA ORDER 4
FORTRAN CODE RKPC

9.3.5 Stiff ODEs

While the Euler, Runge-Kutta, and Predictor-Corrector explicit formulas are useful in a wide class of ordinary differential equations, they are less so in the case of so-called stiff ODEs. There are several possible definitions of what constitutes a "stiff" ODE. Butcher [36] defines stiffness in terms of the spectral resolution of the ODE operator when the eigenvalues scale over many orders of magnitude. Ascher and Petzold [41] define stiffness in terms of the smallness of step size required for stable solution of the Euler method. Hairer et al. [32, 37] simply state that "Stiff equations are problems for which explicit methods don't work". In Chap. IV of volume II these authors give examples of stiff ODEs that arise in chemical reactions, electrical circuits, diffusion and mechanics of elastic beams. These authors (and others) use such examples of stiff ODEs to demonstrate their features and some examples will also be introduced below. The simplest explanation of stiffness is that the solution of a stiff system has components with different time scales, some with very "fast" temporal changes while others are simultaneously "slow" by comparison. It is this combination that requires the judicious choice of step size in fixed step methods (if they are to succeed at all) to avoid the regions of instability inherent in all such methods. Eventually, for stiff problems, explicit single step methods are replaced by implicit multi-step and multi-order methods such as the ground-breaking work on Backward Differentiation Formulas (BDF) by Gear [42]. A simple summary is to be found in the online publication at Scholarpedia [43]. Such methods allow for the step size and order of the formula to be changed as the algorithm progresses based on convergence test criteria, thereby reducing the computational overhead in comparison with single step formulas.

In general, a stiff system of ODEs consists of a sequence of first-order equations such as those used in the accompanying Fortran code RKPC [40]. This has the form shown here with the respective initial values

$$\frac{du(x)}{dx} = Au(x), \tag{9.22}$$

$$u_0 = \begin{bmatrix} 2 \\ 1 \\ 2 \end{bmatrix}, \tag{9.23}$$

with A a 3×3 matrix

$$A = \begin{bmatrix} -0.1 & -49.9 & 0 \\ 0 & -50 & 0 \\ 0 & 70 & -120 \end{bmatrix}, \tag{9.24}$$

where u is a vector with three components having the known solution [38]

$$u_1(x) = e^{-0.1x} + e^{-50x}, \tag{9.25}$$

$$u_2(x) = e^{-50x}, \tag{9.26}$$

$$u_3(x) = e^{-50x} + e^{-120x}. \tag{9.27}$$

The eigenvalues of matrix A are $\lambda_1 = -0.1$, $\lambda_2 = -50$, $\lambda_3 = -120$, and it is clear that they scale over four orders of magnitude. The solution of (9.22) has three temporal scales and any multi-step method starting with the initial value (9.23) must be stable enough to resolve the three rates of temporal decay. The choice of step size needs to follow the fastest change and for explicit multi-step methods this requirement calls for very small values. This becomes clear in the application of the Fortran code RKPC where the stiff system VI of (9.22) required at least an order of magnitude smaller step size compared to the non-stiff cases I, II, III of (9.18).

To demonstrate this for a simple, single stiff ODE, consider the example from Sect. 6.7 of [38]

$$\frac{du}{dx} = -200\,[u - F(x)] + \frac{dF}{dx}, \tag{9.28}$$

$$F(x) = 10 - (10 + x)\,e^{-x}, \tag{9.29}$$

$$\frac{dF}{dx} = (9+x)e^{-x}, \tag{9.30}$$

$$u_0 = 10, \tag{9.31}$$

with the exact solution

$$u(x) = F(x) + 10\, e^{-200x}. \tag{9.32}$$

Here both Runge-Kutta and predictor-corrector formulas of order 4 were applied in the examples. Inspection of the error at small starting values of the argument shows the difficulty that the RK method experiences in resolving the rapid decay at small step sizes. However, the PC method is more stable and rapidly converges to the limit of numerical precision $\sim 10^{-15}$ on a commodity hardware platform, but at the expense of 23968 function evaluations! Note should also be made of the oscillations in the error curve similar to those observed when explicit formulas are applied in the examples of Sect. IV.1 of [37].

The application of Gear's Fortran code to the example of (9.22), (9.23) shows remarkably different performance with variable order and variable step size algorithms for both Adams PC and BDF methods. The modified Fortran code GEAR (from [42]) is used in both methods to produce the summary shown in Examples 9.21 and 9.22, where termination is determined by the input of the number of steps and the required accuracy ($\sim 10^{-6}$). At the bottom of each example is a histogram table summarizing the orders used and the number of steps. Also shown is the number of function calls which is small in comparison with those for non-stiff cases in the previous examples. Parenthetically it is noted that the Gear algorithm is the only solver used in the Global Climate Model (GCM)[2] at NASA for the chemical reaction rate time steps. It is one of three solvers used in the Community Multi-scale Aqueous Chemistry (CMAQ)[3] model developed and applied by the U.S. EPA. For both of these models, the author has developed a sparse matrix solver using a Fortran version FSPARSE of Davis's CSparse library [44] (see Sect. 5.12).

EXAMPLE 9.19: RUNGE- KUTTA ORDER 4
EXAMPLE 9.20: PREDICTOR- CORRECTOR ORDER 4
FORTRAN CODE RKPC
EXAMPLE 9.21: GEAR PC ORDER 4
EXAMPLE 9.22: GEAR ALGORITHM ORDER 4
FORTRAN CODE GEAR

---

[2] https://www.giss.nasa.gov/projects/gcm/.
[3] https://www.epa.gov/cmaq.

9.4 Spectral Function Methods

9.4.1 Spectral Functions

In [23] a spectral function method was developed for the first-order ODE in the following form:

$$\frac{\partial u}{\partial t} = \kappa \frac{\partial^2 u}{\partial r^2} + Q(r, t, u), \tag{9.33}$$

where the left-hand side is a derivative with respect to time and the right-hand side has two derivatives with respect to one spatial degree of freedom. The domain of definition for the solution $u(r, t)$ is $r \in [0, \rho]$ and $t \in [0, \tau]$, with boundary conditions

$$u(0, t) = 0,$$
$$u(\rho, t) = 0, \tag{9.34}$$

and initial condition

$$u(r, 0) = u_0(r). \tag{9.35}$$

The right-hand side of (9.33) contains the source term $Q(r, t, u)$ that depends only on r, in the linear case, or on $u(r, t)$ for the non-linear case. The coefficient κ may be chosen as a constant, or have some other value. For the discussion in this section and in the examples, only the linear case is considered, although the same set of spectral functions was used in both linear and non-linear cases in [23] as is described in Sect. 9.5.2 for a specific non-linear example from [23].

9.4.2 Choice of Basis

The essence of the spectral function method is that of separation of variables, by representation of the solution as a sum of products of functions in the two different degrees of freedom, time t, and physical space $r = \rho x$, $x \in [-1, +1]$

$$u(r, t) = \sum_{n=1}^{\infty} p_n(t) w_n(x). \tag{9.36}$$

The set $\{p_n(t)\}_1^\infty$ depends only on the variable $t \in [0, \tau]$, while $\{w_n(x)\}_1^\infty$ is orthonormal on the interval of x with respect to a weight function $\varphi(x) > 0$ on the same interval. In other words, in the terminology of Sect. 4.5, for a Hilbert space **H**, with an inner product (4.81)

$$(f, g) = \int_{-1}^{+1} f(x) g(x) \varphi(x) dx \tag{9.37}$$

9.4 Spectral Function Methods

ensures that (9.36) is a Fourier series [45] for $u(r,t)$. For a fixed t the solution is fully specified by the set of bounded numbers $\mathbf{p} = \{p_n(t)\}_1^\infty \subset \mathbf{h}$ and the Hilbert spaces $\mathbf{h}$ and $\mathbf{H}$ are isomorphic. But the exact Fourier coefficients are not known *a priori* and usually a Galerkin method is used to obtain approximate values for the exact Fourier coefficients for each value of t. This projects the representation (9.36) onto a subspace $\mathbf{H}_N \subset \mathbf{H}$ with dimension N, and the evaluation of the coefficients is now described.

The choice of basis for the spectral functions is usually limited to trigonometric or polynomial functions, although higher transcendental functions have been used [46]. This discussion follows the development in Sect. B of [23] and is applicable to both the diffusion and wave equations.

Let $\mathbf{t}$ be the vector of $N+1$ components such that, in the trigonometric case, the transposed vector is

$$\mathbf{t}^T = \{cos(0\pi x), sin(1\pi x), cos(1\pi x), sin(2\pi x), cos(2\pi x), \\ \ldots\ldots, sin(N\pi x/2), cos(N\pi x/2)\}, \tag{9.38}$$

and, in the Chebyshev polynomial case,

$$\mathbf{t}^T = \{T_0(x), T_1(x), T_2(x), T_3(x), \ldots, T_N(x)\}, \tag{9.39}$$

where $r = \rho x$, and $x \in [-1,+1]$, with N chosen as an even integer. Then the truncated form of (9.36) is derived from the expansion

$$\mathbf{u} = \mathbf{t}^T \mathbf{a}, \tag{9.40}$$

where $\mathbf{a}$ is the vector of $N+1$ coefficients that depend on t

$$\mathbf{a}^T = \{\frac{1}{2}a_0(t), a_1(t), a_2(t), a_3(t), \ldots, a_N(t)\}. \tag{9.41}$$

The following sections describe how the expression of (9.36) is obtained from this choice of basis functions.

9.4.3 Incorporating Boundary Conditions

The first step in forming the spectral functions is to apply the boundary conditions (9.34) to show that two of the coefficients in (9.41) are linear combinations of the others. In the Chebyshev case, these are coefficients $\frac{1}{2}a_0$, a_1, while in the trigonometric case they are $\frac{1}{2}a_0$, a_2. Substituting the appropriate linear combinations in (9.40) and rearranging terms, a set of linearly independent functions $\{e_n(x)\}_1^{N-1}$ follows and each satisfies both boundary conditions in (9.34).

For the trigonometric case

$$u(r,t) = \sum_{n=1}^{\frac{N}{2}} a_{2n-1}(t)\, e_n(x) +$$

$$\sum_{n=1}^{\frac{N}{2}-1} a_{2n+2}(t)\, e_{\frac{N}{2}+n}(x), \qquad (9.42)$$

where for $n = 1, \ldots, \frac{N}{2}$

$$e_n(x) = sin(n\pi x), \qquad (9.43)$$

and for $n = 1, \ldots, \frac{N}{2} - 1$

$$e_{\frac{N}{2}+n}(x) = cos\{(n+1)\pi x\} - cos(0\pi x), \; n \text{ odd}, \qquad (9.44)$$
$$= cos\{(n+1)\pi x\} - cos(1\pi x), \; n \text{ even}.$$

For the Chebyshev case

$$u(r,t) = \sum_{n=1}^{\frac{N}{2}-1} a_{2n+1}(t)\, e_n(x) +$$

$$\sum_{n=0}^{\frac{N}{2}-1} a_{2n+2}(t)\, e_{\frac{N}{2}+n}(x), \qquad (9.45)$$

where for $n = 1, \ldots, \frac{N}{2} - 1$

$$e_n(x) = T_{2n+1}(x) - T_1(x), \qquad (9.46)$$

and for $n = 0, \ldots, \frac{N}{2} - 1$

$$e_{\frac{N}{2}+n}(x) = T_{2n+2}(x) + T_0(x) - 2T_1(x), \; n \text{ even}, \qquad (9.47)$$
$$= T_{2n+2}(x) - T_0(x), \; n \text{ odd}.$$

Each member of the linearly independent set $\{e_i\}_1^{N-1}$ satisfies the boundary conditions and a matrix transformation defines them in terms of the chosen basis as

$$\mathbf{e} = \mathbf{T}\mathbf{t}, \qquad (9.48)$$

where $\mathbf{e}$ is the vector of $N-1$ coordinate functions $e_n(x)$. The vector $\mathbf{t}$ in (9.39) has the elements $\{t_j(x)\}_1^{N+1}$ and the matrix $\mathbf{T}$ has matrix elements $T_{i,j}$ and is rectangular with $N-1$ rows and $N+1$ columns because $\mathbf{e}$ has two elements less than $\mathbf{t}$. Both are different if the boundary conditions, or basis set is changed. In the following the Chebyshev basis will be described in detail.

9.4.4 Orthonormalization

The set **e** is linearly independent and could be used to solve (9.33), but it is much simpler if an orthonormal set is used instead. For this reason the modified Gram-Schmidt (MGS) method is applied to transform the linearly independent set into an orthonormal set $\{w_n\}_1^{N-1}$ as proposed in (9.36). The following algorithm constructs a matrix **S** to transform the basis vector **t** into the orthonormal set.

For the Chebyshev case the MGS method has the initial steps described as follows, after setting arrays

$$S_{n,i} = 0,$$
$$D_{n,i} = 0, n = 1, \ldots, N-1, i = 1, \ldots, N+1. \tag{9.49}$$

Then, for initialization with $n = 1$, from (9.46) $t_2(x) = T_1(x)$ and $t_4(x) = T_3(x)$, with their corresponding matrix elements of **T**, we have

STEP 1: $\quad e_1(x) = T_{1,2} t_2(x) + T_{1,4} t_4(x),$ \hfill (9.50)

STEP 2: $\quad d_1(x) = e_1(x),$
$$= D_{1,2} t_2(x) + D_{1,4} t_4(x), \tag{9.51}$$
$$D_{1,2} = T_{1,2},$$
$$D_{1,4} = T_{1,4},$$

STEP 3: $\quad \| d_1 \| = (d_1, d_1),$
$$= \sqrt{\frac{\pi}{2} \left(D_{1,2}^2 + D_{1,4}^2 \right)}, \tag{9.52}$$
$$= \sqrt{\pi},$$

STEP 4: $\quad S_{1,2} = \dfrac{D_{1,2}}{\| d_1 \|},$ \hfill (9.53)
$$S_{1,4} = \frac{D_{1,4}}{\| d_1 \|},$$

STEP 5: $\quad w_1(x) = S_{1,2} t_2(x) + S_{1,4} t_4(x).$ \hfill (9.54)

Then, for subsequent steps with $n > 1$, the MGS algorithm for $n = 2, \ldots, N-1$ continues with

STEP 1: $\quad e_n(x) = \displaystyle\sum_{j=1}^{N+1} T_{n,j} t_j(x),$ \hfill (9.55)

STEP 2: $\quad \alpha_{n,i} = -(w_i, e_n), \ i = 1, \cdots, n-1,$

$$= -\frac{\pi}{2}\left[2S_{i,1}T_{n,1} + \sum_{j=2}^{N+1} S_{i,j}T_{n,j}\right], \qquad (9.56)$$

STEP 3: $\quad d_n = e_n(x) + \sum_{i=1}^{n-1} \alpha_{n,i} w_i(x),$

$$= \sum_{i=1}^{N+1} D_{n,i} t_i(x),$$

$$D_{n,i} = T_{n,i} + \sum_{j=1}^{n-1} \alpha_{n,j} S_{j,i},$$

$$\|d_n\| = \sqrt{\frac{\pi}{2}\left(2D_{n,1}^2 + \sum_{i=2}^{N+1} D_{n,i}^2\right)}, \qquad (9.57)$$

STEP 4: $\quad S_{n,i} = \dfrac{D_{n,i}}{\|d_n\|}, \ i = 1, \cdots, N+1, \qquad (9.58)$

STEP 5: $\quad w_n(x) = \sum_{i=1}^{N+1} S_{n,i} t_i(x). \qquad (9.59)$

The MGS algorithm outlined here uses the expressions in Sect. B of [23] to compute the matrix elements of **S** in the definition of the orthonormal set $\{w_n\}_1^{N-1}$ as a transform of the basis **t**

$$\mathbf{w} = \mathbf{S}\mathbf{t}. \qquad (9.60)$$

The solution of (9.33) has the spectral representation of (9.36) in matrix notation

$$\mathbf{u} = \mathbf{w}^T \mathbf{p}. \qquad (9.61)$$

EXAMPLE 9.23: FOR N = 7 COMPUTE MATRICES T AND S
EXAMPLE 9.24: FOR N = 9 COMPUTE MATRICES T AND S

9.4.5 The Second-Order Spatial Derivative

The first-order non-linear diffusion equation (9.33) has the following form, if the inhomogeneous term is neglected to leave a linear equation

$$\frac{\partial u}{\partial t} = \kappa \frac{\partial^2 u}{\partial r^2}, \qquad (9.62)$$

9.4 Spectral Function Methods

where the left-hand side is a derivative with respect to time and the right-hand side has two derivatives with respect to one spatial degree of freedom. Substitution of (9.61) into this linear form of (9.33) gives the linear equation in matrix form

$$\mathbf{w}^T \frac{d\mathbf{p}}{dt} = \frac{\kappa}{\rho^2} \mathbf{w}''^T \mathbf{p}, \qquad (9.63)$$

where $\mathbf{w}''$ is a vector with elements

$$w_n''(x) = \sum_{i=1}^{N+1} S_{n,i} t_i''(x), \qquad (9.64)$$

and the double prime denotes the second derivative with respect to x. This term implies the second-order derivative of the Chebyshev polynomial basis (9.39) and Luke (Sect. 8.5 p. 299 [47]) has given expressions for their expansion in the same polynomials. These are for even powers of x

$$t''_{2m+1}(x) = \sum_{k=0}^{m-1} c_{m,k} t_{2k+1}(x), \quad 1 \le m \le \frac{N}{2}, \qquad (9.65)$$

and on noting that

$$t_{2m+1}(x) = T_{2m}(x), \quad 0 \le m \le \frac{N}{2}, \qquad (9.66)$$

the coefficient in (9.65) is

$$\begin{aligned} c_{m,0} &= 4m^3, \ k=0, \\ c_{m,k} &= 8m(m^2 - k^2), \ 1 < k \le m-1. \end{aligned} \qquad (9.67)$$

While for odd powers of x

$$t''_{2m}(x) = \sum_{k=1}^{m-1} c^o_{m-1,k-1} t_{2k}(x), \quad 2 \le m \le \frac{N}{2}, \qquad (9.68)$$

and on noting that

$$t_{2m}(x) = T_{2m-1}(x), \quad 1 \le m \le \frac{N}{2}, \qquad (9.69)$$

the coefficient in (9.68) is

$$c^o_{m-1,k-1} = 4(2m-1)(m-k)(m+k-1), \ 1 \le k \le m-1. \qquad (9.70)$$

Substitution of (9.65) and (9.68) into (9.64) gives, respectively, for even powers of x

$$w_n''(x) = \sum_{k=0}^{\frac{N}{2}-1} K_{n,k} t_{2k+1}, \qquad (9.71)$$

where the coefficient is

$$K_{n,k} = \sum_{m=k+1}^{\frac{N}{2}} S_{n,2m+1} c_{m,k}, \tag{9.72}$$

while for odd powers of x

$$w_n''(x) = \sum_{k=1}^{\frac{N}{2}-1} K_{n,k}^o t_{2k}, \tag{9.73}$$

where the coefficient is

$$K_{n,k}^o = \sum_{m=k+1}^{\frac{N}{2}} S_{n,2m} c_{m-1,k-1}^o. \tag{9.74}$$

Here (9.72) and (9.74) follow on application of the expansion (9.64). An erratum is noted here for Sect. C of [23] where the lower limit of the summations in expressions (9.72) and (9.74) is $m = k + 1$ as shown here.

The result of all this algebra is that multiplication of (9.63) from the left by the vector **w** leaves the left-hand side with a diagonal matrix having unity on the diagonal. Whereas the right-hand side has a matrix **W** whose elements are obtained from (9.72) and (9.74), by noting the orthogonality relation of Chebyshev polynomials (2.48)

$$(w_{n'}, w_n'') = \frac{\pi}{2} \left\{ 2 S_{n',1} K_{n,0} + \sum_{k=1}^{\frac{N}{2}-1} \left[S_{n',2k+1} K_{n,k} + S_{n',2k} K_{n,k}^o \right] \right\}. \tag{9.75}$$

Thus the diffusion equation in the linear form of (9.63), after left multiplication by the vector **w**, reduces to the vector ODE

$$\frac{d\mathbf{p}}{dt} = \frac{\kappa}{\rho^2} \mathbf{W} \mathbf{p}, \tag{9.76}$$

with the matrix elements of **W** defined by (9.75).

EXAMPLE 9.25: FOR N = 7 EIGENVALUES AND EIGENVECTORS OF W
EXAMPLE 9.26: FOR N = 9 EIGENVALUES AND EIGENVECTORS OF W

9.4.6 Spectral Mappings

The first-order non-linear diffusion equation (or heat equation) was investigated with spectral function methods in [23]. This included a short list of linear cases including P1 and P2 (from Rektorys [23,48]), on the interval $r \in [0, \pi]$, with the following initial conditions:

9.4 Spectral Function Methods

$$\begin{aligned} \text{P1} \quad & u_0(r) = sin(1 \times r), \\ \text{P2} \quad & u_0(r) = sin(2 \times r). \end{aligned} \quad (9.77)$$

These have the known solutions in closed form as

$$u(r,t) = e^{-n^2 t} sin(nr), \; n = 1, 2, \quad (9.78)$$

that may be compared against a numerical solution using the spectral function representation of (9.76). In the Fourier case of (9.38), the matrix $\mathbf{W}$ is diagonal, but not in the Chebyshev case and it may not be symmetric either. In the latter case it is diagonalized as the form

$$\mathbf{W} = \mathbf{Y} \Lambda \mathbf{Z}^T, \quad (9.79)$$

where Λ is the diagonal matrix of (real) eigenvalues of $\mathbf{W}$ arranged in increasing magnitude. The matrices $\mathbf{Y}, \mathbf{Z}$ have columns that are eigenvectors of $\mathbf{W}, \mathbf{W}^T$, respectively, with normalization[4]

$$\mathbf{Z}^T \mathbf{Y} = \mathbf{Y} \mathbf{Z}^T = \mathbf{1}. \quad (9.80)$$

Left multiplication of (9.76) by $\mathbf{Z}^T$ yields the vector equation

$$\frac{d\mathbf{p}^{(z)}}{dt} = \frac{\kappa}{\rho^2} \Lambda \, \mathbf{p}^{(z)}, \quad (9.81)$$

which is the spectral resolution of (9.62), with

$$\mathbf{p}^{(z)} = \mathbf{Z}^T \mathbf{p}, \quad (9.82)$$

hence the name "spectral method". The form (9.79) implies a solution of a rank N matrix eigenvalue problem for eigenvalues and eigenvectors. While a rank N system will have N eigenvalues and eigenvectors, not all will have converged, as is demonstrated in the EXAMPLES 9.25,9.26 for N = 7 and 9, respectively. In [23] Figs. 3 and 4 show, respectively, the eigenvalue magnitude and how convergence improves for each, with increasing N and in particular the correlation between N and the number of digits of precision. In the following discussion only the N = 9 case will be used for demonstration purposes, but note should be taken that less accurate values for eigenvalues and corresponding eigenvectors are to be expected with such small values of N especially above the lowest eigensolutions.

Thus (9.81) is a vector equation in frequency (or spectral) space and the back transform to physical (coordinate) space is

$$\mathbf{p} = \mathbf{Y} \mathbf{p}^{(z)}. \quad (9.83)$$

The eigenvalues Λ of the exact solution are known to be $(n\pi)^2$ in ascending order, or n^2 for $\mathbf{W}/\rho^2$, when $\rho = \pi$.

---

[4] Compare this to (5.56).

The mappings described so far are not as useful as those that map to and from the Chebyshev basis of (9.39) because of their favorable properties. For this purpose two such mappings are required

$$R : \{a_i\}_0^N \rightarrow \{p_i\}_1^{N-1}, \tag{9.84}$$

$$R^{-1} : \{p_i\}_1^{N-1} \rightarrow \{a_i\}_0^N, \tag{9.85}$$

and matrix representations follow on comparing (9.36), (9.40) and (9.41)

$$\mathbf{w}^T \mathbf{p} = \mathbf{t}^T \mathbf{a}, \tag{9.86}$$

where left multiplication by vector $\mathbf{w}$ or $\mathbf{t}$, followed by application of the inner product (2.48), yields, respectively:

$$\mathbf{p} = \mathbf{R}\,\mathbf{a}, \tag{9.87}$$

$$\mathbf{R}^{-1}\mathbf{p} = \mathbf{a}. \tag{9.88}$$

It follows from (9.60) and the definition (9.41) that in the Chebyshev case $\mathbf{R} = (\mathbf{w}, \mathbf{t}^T)$ is $\frac{\pi}{2}$ times the matrix $\mathbf{S}$ with the first column multiplied by 2. In the Fourier case this multiplier is absent. Similarly, $\mathbf{R}^{-1} = \mathbf{S}^T$ for both Fourier and Chebyshev cases. It should be noted that matrix $\mathbf{R}$ has $N-1$ rows and $N+1$ columns, while the converse is the case for $\mathbf{R}^{-1}$.

9.4.7 A Case Study of the Non-linear Diffusion Equation

The vector equation (9.81) may now be integrated in spectral space by any of the standard RK or PC methods described in Sect. 9.3. In the linear case with a non-linear term absent, the solution for each eigenmode evolves in time t independent of the other modes. In the non-linear case, a non-linear term may mix the eigenmode solutions and its time dependence must be computed at each time step. It is assumed that the non-linear term in (9.33) has an expansion similar to (9.36) in truncated form

$$Q(r, t, u) = \sum_{n=1}^{N-1} q_n(t)\, w_n(x), \tag{9.89}$$

where the expansion is obtained by a basis transformation similar to (9.40) and (9.86)

$$\mathbf{w}^T \mathbf{q} = \mathbf{t}^T \mathbf{f}, \tag{9.90}$$

where

$$\mathbf{f}^T = \{\frac{1}{2} f_0, f_1, f_2, f_3, \ldots, f_N\} \tag{9.91}$$

is the vector of expansion coefficients of $Q(r, t, u)$ on the basis $\{t_i\}_1^{N+1}$.

9.5 Double Spectral Function Method

If the non-linear term contains simple integer powers of the solution $u(r, t)$, such as $|u(r, t)|^2$, then the algorithm at each time step has three parts:

1. back transform of $\mathbf{p}^{(z)} \to \mathbf{a}$;
2. convolution of the Chebyshev series to form the Chebyshev series $\mathbf{t}^T\mathbf{f} = \mathbf{t}^T\mathbf{a} \otimes \mathbf{t}^T\mathbf{a}$ (as in Sect. 2.9); and
3. forward transform $\mathbf{f} \to \mathbf{q}^{(z)}$ for an added vector $\mathbf{q}^{(z)}$ on the right-hand side of (9.81) followed by
4. advance to the next time step in spectral space.

The details are given in [23]. The examples in this section are for the linear ODE of (9.81) with cases in (9.77) having exact solutions (9.78) where, for simplicity, the coefficient is taken as $\kappa = 1$. In the EXAMPLES the spectral function expansion coefficients are integrated in time with both Euler and Runge-Kutta methods.

EXAMPLE 9.27: EULER'S METHOD FOR FIRST EIGENVALUE
EXAMPLE 9.28: EULER'S METHOD FOR SECOND EIGENVALUE
EXAMPLE 9.29: RUNGE- KUTTA METHOD FOR FIRST EIGENVALUE
EXAMPLE 9.30: RUNGE- KUTTA METHOD FOR SECOND EIGENVALUE

9.5 Double Spectral Function Method

The use of spectral functions (9.60) in physical space and the spectral resolution (9.82) in frequency (or spectral) space for solution of the diffusion equation (9.1) comprise the single spectral function method (SSFM). A similar spectral resolution in the time degree of freedom is introduced here as the double spectral method (DSFM). Because time is on the positive interval only, $t \in [0, \Upsilon]$, $\Upsilon > 0$, it is appropriate to use as a basis the shifted Chebyshev polynomials with argument $y = t/\tau, t \in [0, \tau], y \in [0, 1]$, where the shifted Chebyshev polynomials have the same orthogonality property (2.48) but with the weight function $1/\sqrt{x(1-x)}$

$$\mathbf{t}^{*T} = \{T_0^*(y), T_1^*(y), T_2^*(y), T_3^*(y), \ldots, T_M^*(y)\}, \qquad (9.92)$$

where $T_m^*(y) = T_m(2y - 1)$, with m integer. The vector $\mathbf{a}$ is the vector of $M + 1$ coefficients (9.41) that depend on t, written in matrix form as

$$\mathbf{a} = A\mathbf{t}^*, \qquad (9.93)$$

where for each coefficient in the vector $\mathbf{a}$

$$a_n(t) = \sum_{m=0}^{M\prime} A_{n,m} T_m^*(y). \qquad (9.94)$$

If the inhomogeneous term is present, then it also has the spectral representation in (9.89), with a basis transformation similar to (9.90) for coefficients. If the coefficients $\mathbf{f}$ of the inhomogeneous term are time dependent, then they also have a basis of shifted Chebyshev polynomials similar to (9.93), in the matrix form

$$\mathbf{f} = \mathbf{E}\mathbf{t}^*, \qquad (9.95)$$

where for each coefficient in the vector $\mathbf{f}$

$$f_n(t) = \sum_{m=0}^{M\prime} E_{n,m} T_m^*(y). \qquad (9.96)$$

The specific values of the matrix elements in (9.95) depend on the functional form of the inhomogeneous term $Q(r, t, u)$. Note that the first column of matrix $\mathbf{A}$ (or $\mathbf{E}$) has matrix elements $\frac{1}{2}A_{n0}$ (or $\frac{1}{2}E_{n0}$) for the first shifted Chebyshev series term of $a_n(t)$ (or $f_n(t)$), $n = 0, 1, \cdots, N$, and the matrices have $M+1$ columns and $N+1$ rows. The form for the linear case in (9.76) is augmented by a non-linear term that follows from (9.90). Left multiplication by $\mathbf{w}$, followed by application of the orthonormal property of the spectral function basis gives the result $(\mathbf{w},\mathbf{w}^T)\,\mathbf{q} = \mathbf{q}$. This leads to the non-linear form of (9.76) as

$$\frac{d\mathbf{p}}{dt} = \frac{\kappa}{\rho^2}\,\mathbf{W}\,\mathbf{p} + \mathbf{q}, \qquad (9.97)$$

where the coefficients in vectors $\mathbf{p}, \mathbf{q}$, depend on time. Following the spectral resolution of (9.76), as discussed in Sect. 9.4.6, the vector equation (9.81) becomes

$$\frac{d\mathbf{p}^{(\mathbf{z})}}{dt} = \frac{\kappa}{\rho^2}\,\Lambda\,\mathbf{p}^{(\mathbf{z})} + \mathbf{q}^{(\mathbf{z})}, \qquad (9.98)$$

where analogously to (9.82)

$$\mathbf{q}^{(\mathbf{z})} = \mathbf{Z}^T\,\mathbf{q}, \qquad (9.99)$$

and for simplicity assume $\kappa = 1$ in the following discussion.

9.5.1 The Non-linear Diffusion Equation Solved by Time Stepping

The vector equation (9.98) is stepped forward in time as before but the inhomogeneous term needs to be evaluated in physical space. This requires the back transform of (9.83) and evaluation of the solution from (9.36). In detail this transform from spectral to physical space has two stages for each eigenvector, first from spectral space to physical space $\mathbf{p}^{(\mathbf{z})} \to \mathbf{p}$ followed by recovery of the Chebyshev coefficients, $\mathbf{p} \to \mathbf{a}$. These transforms begin with (9.82), and are followed by (9.87),

9.5 Double Spectral Function Method

(9.93)

$$\begin{aligned} \mathbf{p}^{(z)} &= \mathbf{Z}^T \mathbf{p} \\ &= \mathbf{Z}^T \mathbf{R}\mathbf{a} \\ &= \mathbf{F}\mathbf{a} \\ &= \mathbf{F}\mathbf{A}\mathbf{t}^* \\ &= \mathbf{P}\mathbf{t}^*, \end{aligned} \quad (9.100)$$

using the definitions

$$\mathbf{P} = \mathbf{F}\mathbf{A}, \quad (9.101)$$

$$\begin{aligned} \mathbf{F} &= \mathbf{Z}^T \mathbf{R} \\ &= \frac{\pi}{2}\mathbf{Z}^T \mathbf{S}, \end{aligned} \quad (9.102)$$

where the latter follows after application of (9.87).

Similarly, for the inhomogeneous term in (9.99), after application of (9.95)

$$\begin{aligned} \mathbf{q}^{(z)} &= \mathbf{Z}^T \mathbf{q} \\ &= \mathbf{F}\mathbf{f} \\ &= \mathbf{F}\mathbf{E}\mathbf{t}^* \\ &= \mathbf{Q}\mathbf{t}^*, \end{aligned} \quad (9.103)$$

with the definition

$$\mathbf{Q} = \mathbf{F}\mathbf{E}. \quad (9.104)$$

The inverse transforms of (9.101) and (9.104), are, respectively,

$$\mathbf{F}^{-1}\mathbf{P} = \mathbf{A}, \quad (9.105)$$

$$\mathbf{F}^{-1}\mathbf{Q} = \mathbf{E}. \quad (9.106)$$

For an individual vector component (9.105)

$$\begin{aligned} \mathbf{a} &= \mathbf{F}^{-1}\mathbf{p}^{(z)} \\ &= \mathbf{F}^{-1}\mathbf{P}\mathbf{t}^* \\ &= \mathbf{A}\mathbf{t}^*, \end{aligned} \quad (9.107)$$

and likewise for the inhomogeneous vector component

$$\begin{aligned} \mathbf{f} &= \mathbf{F}^{-1}\mathbf{q}^{(z)} \\ &= \mathbf{F}^{-1}\mathbf{Q}\mathbf{t}^* \\ &= \mathbf{E}\mathbf{t}^*. \end{aligned} \quad (9.108)$$

If the inhomogeneous term depends on the solution $u(r, t)$ then it must be evaluated at the current time step. When $Q(r, t, u)$ involves integer powers of $u(r, t)$, as is the case for problems P5 and P6 in [23,48], the additional mapping of (9.88), with $\mathbf{R}^{-1} = \mathbf{S}^{\mathbf{T}}$, is required to retrieve the Chebyshev coefficients. Then a convolution of Chebyshev series is performed as indicated in step 2 of Sect. 9.4.7, using the algorithm discussed in Sect. 2.9.

In the simplest instance, consider the case of problem P4 in [23,48] where the inhomogeneous term does not depend on time and has the form

$$Q(r) = sin(r), \ r \in [0, \pi]. \tag{9.109}$$

In this case the exact solution has this initial value and form

$$u(r, 0) = 0,$$
$$u(r, t) = (1 - e^{-t}) sin(r). \tag{9.110}$$

The application of a simple one-step Euler method is shown in EXAMPLE 9.31 and the result is compared with the exact solution (9.110). The difference between the Euler prediction and exact solution (9.110) is due to the relatively small number of spectral functions $N = 9$, but it is noteworthy that the relative error does not grow as time increases. Note that the solution of the inhomogeneous case is proportional to the first eigenmode and grows with time, whereas the solution of the linear case without the inhomogeneous term decays with time.

EXAMPLE 9.31: EULER'S METHOD FOR THE INHOMOGENEOUS TERM

9.5.2 The Non-linear Diffusion Equation Solved by Recurrence

In the previous sections, time stepping algorithms were applied to solve (9.33) because of the derivative in time on the left-hand side. However, in the DSFM this is also solved by use of a Chebyshev expansion of the time-dependent coefficients (9.94) from which a three term recurrence for these coefficients is derived. Consider again the spectral form (9.98) for eigenmode component n (note that in this section this is the index for the eigenvalue and eigenvector)

$$\frac{dp_n^{(z)}}{dt} = \lambda_n p_n^{(z)} + q^{(z)}, \tag{9.111}$$

where λ_n is the nth (real) eigenvalue of matrix $\mathbf{W}$ in (9.79). After the change in variable $t = \tau y$ this becomes

$$\frac{dp_n^{(z)}}{d\tau} = \tau \lambda_n p_n^{(z)} + \tau q^{(z)}, \tag{9.112}$$

for which the following expansions in shifted Chebyshev polynomials are introduced

$$p_n^{(z)}(y) = \sum_{m=0}^{M\prime} P_{n,m} T_m^*(y), \tag{9.113}$$

$$\frac{dp_n^{(z)}}{dy} = \sum_{m=0}^{M\prime} P'_{n,m} T_m^*(y), \tag{9.114}$$

9.5 Double Spectral Function Method

where the prime on the summation denotes that the first term contains the usual factor of $\frac{1}{2}$ and (9.113) is seen to be the n component of (9.100). Likewise, for the n component of the inhomogeneous term in (9.103), a corresponding Chebyshev expansion is

$$q_n^{(z)}(y) = \sum_{m=0}^{M\,\prime} Q_{n,m} T_m^*(y), \tag{9.115}$$

and on substituting (9.113) to (9.115) into (9.112) and applying orthogonality of the Chebyshev polynomials, it follows that

$$P'_{n,m} = \tau \lambda_n P_{n,m} + \tau Q_{n,m}. \tag{9.116}$$

From the property found by Clenshaw [49] there is a simple relationship between coefficients of (9.113), (9.114) in the form

$$4m P_{n,m} = P'_{n,m-1} - P'_{n,m+1},$$

which, on application to (9.116), yields a three term recurrence in order m

$$P_{n,m-1} = \frac{4m}{\tau \lambda_n} P_{n,m} + P_{n,m+1} + \frac{1}{\lambda_n} \left(Q_{n,m+1} - Q_{n,m-1} \right). \tag{9.117}$$

This recurrence may be started at sufficiently large order M such that $P_{n,M+1} = 0$, $P_{n,M} = 1$ with the coefficients $Q_{n,m}$ determined by the form of the inhomogeneous term. The results of the recurrence are normalized by the initial condition at $t = 0$, firstly for the case that the inhomogeneous coefficients $Q_{n,m}$ in (9.117) are zero

$$p_n^{(z)}(0) = \sum_{m=0}^{M\,\prime} (-1)^m P_{n,m}$$

which follows on noting the property $T_m^*(0) = (-1)^m$. Again the first term is $\frac{1}{2} P_{n,0}$. In the EXAMPLES it is assumed for simplicity that $\tau = 1$.

The two cases in EXAMPLES 9.32, 9.33 demonstrate the DSFM for the linear form (without the inhomogeneous term) and compare the result with the RK order 3 time stepping method in EXAMPLES 9.29, 9.30. The DSFM for the case with the inhomogeneous term of (9.109) included is shown in EXAMPLE 9.34 which has only one non-zero coefficient $Q_{n,0} = 2$. This is added after the $P_{n,m}$ are normalized. Only the first coefficient in (9.115) is non-zero because of the initial condition (9.110) at $t = 0$. The form (9.110) has a solution that is the inhomogeneous term (9.109) minus the solution of P1 in (9.77) for the first eigenmode of the linear case. Therefore this negative sign is included in the normalization step. Again, the residual errors in the comparison of DSFM against the Runge-Kutta time stepping algorithm are due to the relatively small number of spectral functions used.

The DSFM is also applicable when the inhomogeneous term (9.33) is non-linear, as in (P6 from [23], and Chap. 6 of [48])

$$Q(r, t, u) = \int_0^t \left[1 - u^2(r, t')\right] dt', \quad r \in [0, 1], \tag{9.118}$$

$$u_0(r) = 0. \tag{9.119}$$

This also has a spectral resolution of the form (9.89) and solution using the DSFM involves convolution of series (see Sect. 2.9) and a quadrature formula for the integral (see Chap. 7). The algorithm for the DSFM is a modification of that in Sect. 9.4.7, where at each time step, the matrices in (9.93) and (9.95), respectively, are computed after a back transform from spectral space to physical space where the convolution of two series and the integral is performed. This is followed by a forward transform into spectral space where a new matrix $P_{n,m}$ in (9.113) is computed by recurrence.

The algorithm at each time step has five parts:

1. Compute matrix **P** by downward recurrence of (9.117).
2. Back transform of $\mathbf{A} = \mathbf{F}^{-1}\mathbf{P}$ of (9.105).
3. Compute coefficient matrix **E** of (9.106).
4. Forward transform $\mathbf{Q} = \mathbf{F}\mathbf{E}$ of (9.104).
5. Compute new matrix **P** by downward recurrence of (9.117).

Step 1 provides an initial matrix **P** and the iteration consists of repetition of steps 2 to 5 until the matrix **A** has converged. Only step 3 is dependent on the specific form of the non-linearity in the inhomogeneous term. For the form in (9.118), after back transformation in step 2, matrix **A** is known. The evaluation of **E** requires a convolution of two Chebyshev series and evaluation of the integral as follows (where the use of the index n, now refers to the terms of a series). If the convolution product for u^2 has been evaluated in a Chebyshev series with index n, then it has series coefficients defined as

$$u^2(r, t') = \sum_{n=0}^{N\,\prime} b_n(t')\, t_{n+1}(x) \tag{9.120}$$

and

$$Q(r, t) = t - \sum_{n=0}^{N\,\prime} h_n(t)\, t_{n+1}(x) \tag{9.121}$$

where, for the basis (9.39),

$$h_n(t) = \int_0^t b_n(t') dt'. \tag{9.122}$$

9.5 Double Spectral Function Method

Therefore the coefficients **f** of (9.91) are

$$f_n(t) = 2t\delta_{n,0} - h_n(t), \tag{9.123}$$

where $\delta_{n,0}$ is the Kronecker delta function and the factor of 2 in the first term arises because of the usual factor of $\frac{1}{2}$ in the Chebyshev series. If the coefficients of the convolution result are expanded in shifted Chebyshev polynomials, then

$$b_n(t') = \sum_{m=0}^{M'} B_{n,m} T_m^*(y), \tag{9.124}$$

for $y \in [0, 1]$, and $h_n(t)$ has the expansion

$$h_n(t) = \sum_{m=0}^{M'} H_{n,m} T_m^*(y). \tag{9.125}$$

Elliot [50] (see also pp. 59–60 of [51]) has given expressions relating the Chebyshev expansion coefficients of the integral of a function to those for the function itself. Applying these results to the present case gives, for $m = 0$

$$H_{n,0} = \frac{1}{2}\tau \left[B_{n,0} - \frac{1}{2}B_{n,1} - 2\sum_{k=2}^{M} \left\{ \frac{(-1)^k}{k^2 - 1} \right\} B_{n,k} \right], \tag{9.126}$$

and for $m > 0$

$$H_{n,m} = \frac{\tau}{4m}\left[B_{n,m-1} - B_{n,m+1}\right]. \tag{9.127}$$

The first term in (9.121) has the first two expansion coefficients as 2τ and τ, respectively, therefore, for each n the matrix **E** of (9.95) is defined by

$$\begin{aligned} E_{n,0} &= 2\tau - H_{n,0}, \\ E_{n,1} &= \tau - H_{n,1}, \\ E_{n,m} &= -H_{n,m}, \quad m > 1. \end{aligned} \tag{9.128}$$

Rektorys (see Chap. 6 of [48]) has provided a basic Fortran code for a numerical solution of the diffusion equation with the non-linear term (9.118) in the form

$$u(r, t_j) = b_j \sin(\pi x) + c_j \sin(3\pi x) \tag{9.129}$$

where the values of the time variable are tabulated in steps $j = 0, \cdots, J$ for some maximum integer value J and stepsize increment h. This has been adapted in EXAMPLE 9.35 where coefficients b_j, c_j are tabulated. Note the erratum to the table in [48]: the first non-zero value is for $j = 2$. Note that the solution (9.129) is a combination of the first and third eigenmode solutions of the linear case (9.78) and therefore it is clear that this form of non-linearity will combine two eigenmodes of (9.81).

EXAMPLE 9.32: DSFM AND RK FOR THE FIRST EIGENVALUE
EXAMPLE 9.33: DSFM AND RK FOR THE SECOND EIGENVALUE

EXAMPLE 9.34: DSFM AND RK FOR P4
EXAMPLE 9.35: NUMERICAL SOLUTION FOR P6

9.6 Second-Order ODEs

A second form of the operator equation has the differential operator in time as second order and in the simplest form it is written as

$$\frac{\partial^2 y(r,t)}{\partial t^2} = s^2 \frac{\partial^2 y(r,t)}{\partial r^2}. \tag{9.130}$$

This form of the wave equation is also known as that of a string of density ϱ under tension T, extending from $r = 0$ to $r = L$, with $y(r,t)$ representing a small displacement with $s^2 = T/\varrho$ (see Chap. 3 of [52]). Here s represents the transverse velocity to the wave propagation and $y(r,t)$ will satisfy some boundary conditions with an initial value (IV) depending on the application. A discussion and derivation of this equation in its various forms is to be found in Chap. 2 of [53] where we are told that, if $y(r,t)$ is the displacement of the string from equilibrium at any point at any time, then (9.130) states that the transverse acceleration of any part of the string is proportional to the curvature of that part. There are several assumptions in the derivations including that the string is flexible and has negligible stiffness, has small displacements (relative to the length), with a uniform distribution of mass along the string, no effects of gravity, and no damping from the surrounding medium. However, the last may be added through a term on the right-hand side.

The solution of (9.130) usually proceeds by expressing $y(r,t)$ as a product of two functions in the respective variables

$$y(r,t) = u(r)v(t), \tag{9.131}$$

then substituting into (9.130), dividing both sides by the right-hand side of (9.131) to give

$$\frac{1}{v(t)} \frac{d^2 v(t)}{dt^2} = \frac{s^2}{u(r)} \frac{d^2 u(r)}{dr^2}, \tag{9.132}$$

where the left-hand side depends only on t, and the right-hand side only on r. Both sides must be equal for all r, t but this is only true if they equal the same constant, say, $-\omega^2$. Thus the right-hand side is written as

$$\frac{s^2}{u(r)} \frac{d^2 u(r)}{dr^2} = -\omega^2, \tag{9.133}$$

$$\frac{d^2 u(r)}{dr^2} + k^2 u(r) = 0, \tag{9.134}$$

$$k = \frac{\omega}{s}. \tag{9.135}$$

9.6 Second-Order ODEs

On application of boundary conditions, such as those for a string fixed at both ends

$$u(0) = 0,$$
$$u(L) = 0, \qquad (9.136)$$

when only discrete values of k are allowed (see Sect. C.2.9(a) of [54])

$$k_n = \frac{n\pi}{L}, \qquad (9.137)$$

$$u_n(r) = \sqrt{\frac{2}{L}} \sin(k_n r), \qquad (9.138)$$

$$n = 1, 2, 3, \cdots, \qquad (9.139)$$

where $k_n, u_n(r)$ are, respectively, the eigenvalues and eigenfunctions of (9.130).

The left-hand side of (9.132) has the general solution (see Sect. C.2.9(a) of [54])

$$v_n(t) = [A_n \sin(\omega_n t) + B_n \cos(\omega_n t)],$$

so that the general solution satisfying (9.130), with these boundary conditions, at all times t, is

$$y(r, t) = \sum_{n=1}^{\infty} y_n(r, t),$$

$$y_n(r, t) = [A_n \sin(\omega_n t) + B_n \cos(\omega_n t)] u_n(r), \qquad (9.140)$$

and

$$\omega_n = s k_n. \qquad (9.141)$$

The values of constants A_n, B_n depend on the initial value, shape of the distribution and transverse speed s, on $r \in [0, L]$. In addition, the initial conditions need to specify the velocity distribution of the displacement in (9.140)

$$\frac{\partial y(r, t)}{\partial t} = \sum_{n=1}^{\infty} \frac{\partial y_n(r, t)}{\partial t},$$

$$= \sum_{n=1}^{\infty} \omega_n [A_n \cos(\omega_n t) - B_n \sin(\omega_n t)] u_n(r). \qquad (9.142)$$

Thus the initial values are

$$y(r, 0) = \sum_{n=1}^{\infty} B_n u_n(r) \qquad (9.143)$$

$$\frac{\partial y(r, t)}{\partial t}\bigg|_{t=0} = \sum_{n=1}^{\infty} A_n \omega_n u_n(r) \qquad (9.144)$$

$$= 0. \qquad (9.145)$$

The functions $\{u_n(r)\}_1^\infty$ form an orthonormal set on the interval $r \in [0, L]$

$$\int_0^L u_n(r) u_m(r) dr = 0, \quad n \neq m,$$

$$= 1, \quad n = m. \quad (9.146)$$

The coefficients A_n, B_n are determined from the initial conditions through the expressions

$$A_n = \frac{1}{\omega_n} \int_0^L u_n(r) \frac{\partial y(r, t)}{\partial t}\Big|_{t=0} dr, \quad (9.147)$$

$$B_n = \int_0^L u_n(r) y(r, 0) dr. \quad (9.148)$$

The expression for B_n follows on multiplying (9.140) with $n = m$ and applying (9.146) and similarly for A_n.

The ω_n are eigenvalues and the y_n the eigenfunctions, or normal modes, of the system (see EXAMPLE 5.1). The eigenvalues are known as the frequencies of the system and ω_1 is the fundamental frequency with all higher harmonics as multiples of the fundamental

$$\omega_1 = \frac{\pi s}{L}, \quad (9.149)$$

$$\omega_n = n \omega_1. \quad (9.150)$$

The eigenmodes remain fixed and distinct for all time, as in vibrations of a string fixed at both ends. Variations on (9.130) may include a forcing term $F(r, t)$ on the right-hand side, or a free boundary at $r = L$. In the latter case, a wave shape may propagate in time along the string. A wave of this type is a displacement of the string from equilibrium as a function of $\xi = r - st$, for a wave traveling in the positive direction and similarly with $r + st$ for a wave traveling in the opposite direction. A standing wave, such as a string fixed at both ends, may be viewed as a superposition of two such traveling waves. The right-hand side of (9.130) reduces to the same spectral resolution as that in Sect. 9.4.6 because the boundary condition (9.136) is the same as (9.34) if $L = \pi$. Because of this the same spectral functions are used in the examples for the wave equation.

Given boundary values and initial conditions, the time integration of (9.130) requires a time stepping algorithm. For this purpose a numerical method for a second-order ODE (see Sect. II.13 of [32]) due to Nyström was used with coefficients for an order 4 formula shown in Tables 9.7 and 9.8 using the template of Table 9.1 with an additional row. These simple formulas are for the case that the right-hand side of (9.130) does not depend on the time derivative of $y(r, t)$ (see Table 13.2 of [32]).

9.6 Second-Order ODEs

Table 9.7 Symbols

		Order	4	
	0			
c_2		$\bar{a}_{21}$		
c_3		$\bar{a}_{31}$	$\bar{a}_{32}$	
$\bar{w}_i$		$\bar{w}_1$	$\bar{w}_2$	$\bar{w}_3$
w_i		w_1	w_2	w_3

Table 9.8 Coefficients

		Values		
	0			
$\frac{1}{2}$		$\frac{1}{8}$		
1		0	$\frac{1}{2}$	
$\bar{w}_i$		$\frac{1}{6}$	$\frac{1}{3}$	0
w_i		$\frac{1}{6}$	$\frac{4}{6}$	$\frac{1}{6}$

On introducing the notation

$$y_j = y(r, j \times h),$$
$$y'_j = \frac{\partial y_n(r,t)}{\partial t}\big|_{t=j\times h}, \qquad (9.151)$$
$$j = 0, 1, 2, \cdots,$$

the next time step is advanced by h using the sequence

$$\begin{aligned}
k'_1 &= f(x_j, y_j), \\
k'_2 &= f(x_j + c_2 h, y_j + c_2 h y'_j + h^2 \bar{a}_{21} k'_1), \\
k'_3 &= f(x_j + c_3 h, y_j + c_3 h y'_j + h^2 \bar{a}_{31} k'_1 + h^2 \bar{a}_{32} k'_2), \\
y_{j+1} &= y_j + h y'_j + h^2 \sum_{i=1}^{3} \bar{w}_i k'_i, \\
y'_{j+1} &= y'_j + h \sum_{i=1}^{3} w_i k'_i.
\end{aligned} \qquad (9.152)$$

This shows the explicit expression, but in the simple examples considered here, the argument is for x where $r = Lx = \rho x = \pi x$, as in the diffusion equation example and does not appear in the right-hand side expression of (9.130).

EXAMPLE 9.36 WAVE EQUATION WITH THE FIRST EIGENMODE
EXAMPLE 9.37 WAVE EQUATION WITH THE SECOND EIGENMODE

9.7 Exercises

1. Hairer and Wanner in Sect. IV.1 of [37] open with a simple example of a stiff 1-D ODE studied by Curtis and Hirschfelder [55]

$$u' = -50[u - \cos(x)]$$

where a possible solution close to $y \approx \cos(x)$ is reached after a transient phase typical of stiff equations. A simple Euler method iterates this as

$$u_{j+1} = u_j - h \times 50[u_j - \cos(x_j)], \quad j = 0, 1, 2, \cdots$$

   (a) For an initial value of $u_0 = 0$ at $x_0 = 0$ apply this Euler method with a step size of $h = \frac{1.974}{50}$ for the first few steps $j = 0, 1, 2, 3, 4$, and tabulate the corresponding values of u_j. Comment on what you observe.
   (b) Repeat the exercise of (a) with a step size of $h = \frac{1.874}{50}$ and compare the two tables of values for u_j.
   (c) Repeat the exercise of (a) with the initial value of $u_0 = \frac{2500}{2501}$.
   (d) Repeat the exercise of (b) with the initial value of (c), compare all four tables of u_j and comment on what you observe.

1. For the Euler method of Exercise 1(a)

   (a) repeat the iteration for a step size of $h = \frac{2.0}{50}$;
   (b) repeat the iteration for a step size of $h = \frac{2.2}{50}$;
   (c) compare all the tabulations of u_j for both exercises and comment on what you observe in the progressive changes in value.

1. In Sect. 9.4.7 the expansion of (9.89) suggests a representation of the inhomogeneous term in a spectral function basis.

   (a) If $Q(r, t, u) = \sin(r)$ where $r \in [0, \pi]$, obtain the expression for the coefficients (9.91) on the Chebyshev basis of (9.39) from the exact values of EXAMPLE 2.37.
   (b) Obtain the coefficients **q** from the transform (9.90) by following the discussion after (9.86).

9.8 Programming Problems

1. Write a code to apply the Euler method described in Exercise 1 and use it for the cases 1(a) to (d) for the results in the first few steps to debug the code. Apply

9.8 Programming Problems

the code to repeat Exercise 1(a) to 1(d) for steps $j = 0, 1, 2, \cdots, 40$. Using $u(x) \approx \cos(x)$ as a crude approximation to the smooth solution, inspect the difference between this and the Euler prediction tabulated at each step in argument x. Plot the results of the Euler solution and the error on the same interval. Compare all cases and comment on what you observe.

2. Using the algorithm of EXAMPLE 9.7 write a code to apply the Runge-Kutta order three formula from Sect. 9.3.3 for the coefficients of Table 9.3 in (9.16) and

$$f(x, u) = -50[u - \cos(x)]$$

with $h = \frac{2.0}{50}$ and $u_0 = \frac{2500}{2501}$, using steps $j = 0, 1, 2, \cdots, 80$. Note that in this case the steps in argument x now appear explicitly in $f(x, u)$ at each step in forming the multipliers r_1, r_2, r_3. Using $u(x) \approx \cos(x)$ as a crude approximation to the smooth solution, inspect the difference between this and the RK prediction tabulated at each step in argument x. Plot the results of the RK solution and the error on the same interval. Compare to the Euler predictions in the previous programming problem and comment on what you observe.

3. Inspect EXAMPLES 9.32, 9.33, 9.34 and repeat them with more spectral functions.

4. As a challenge, consider writing a code that applies the DSFM to non-linear diffusion problem P5 from [23] where

$$Q(r, t, u) = -20\, u(x, t)^3, \quad x \in [0, 1], \ t \in [0, 1]$$

with initial condition

$$u(x, 0) = u_0(x),$$

and boundary conditions

$$u(0, t) = 0$$
$$u(1, t) = 0$$

where

$$u_0(x) = x, \quad 0 \leq x \leq \frac{1}{2}$$
$$= 1 - x \text{ for } \frac{1}{2} \leq x \leq 1.$$

This problem was described in Chap. 5 of [48] and has a numerical solution that may be generated from the Fortran program found in that reference. This numerical solution may be used as a comparison for the DSFM solution. Note that in this case because the solution has discontinuities, the number of spectral functions will need to be large.

References

1. Taylor ME (2022) Introduction to differential equations, vol 52, 2nd edn. Pure and applied undergraduate texts. American Mathematical Society, New York
2. Evans LC (2010) Partial differential equations, vol 19, 2nd edn. Graduate studies in mathematics. American Mathematical Society, New York
3. Rustum C (2022) Partial differential equations a first course, vol 54. Pure and applied undergraduate texts. American Mathematical Society, New York
4. Holt M (1982) Numerical methods in Fluid Dynamics, 2nd edn. Springer-Verlag, Berlin
5. Richtmyer RD, Morton KW (1967) Difference methods for initial-value problems, 2nd edn. John Wiley and Sons, New York
6. Fox L (1990) The numerical solution of two-point boundary problems. Dover Publications, New York, NY
7. Cheung YK, Yeo MF (1979) A practical introduction to finite element analysis. Pitman Publishing Limited, London
8. Davies AJ (1980) The finite element method: a first approach. Oxford University Press, Oxford, UK
9. Girault V, Raviart P-A (1986) Finite element methods for Navier-Stokes equations. Springer-Verlag, Berlin
10. Fletcher C (1991) Computational techniques for fluid dynamics I, 2nd edn. Springer-Verlag, Berlin
11. Fletcher C (1991) Computational techniques for fluid dynamics II, 2nd edn. Springer-Verlag, Berlin
12. Späth H (1995) One dimensional spline interpolation algorithms. A.K. Peters Ltd, Wellesley, MA
13. Späth H (1995) Two dimensional spline interpolation algorithms. A.K. Peters Ltd, Wellesley, MA
14. Birkhoff G, Lynch RE (1984) Numerical solution of elliptical problems. Society for Industrial and Applied Mathematics, Philadelphia, PA
15. Duffy DG (1986) Solution of partial differential differential equations. TAB Books Inc, Blue Ridge Summit PA
16. Gladwell I, Wait R (eds) (1979) A survey of numerical methods for partial differential differential equations. Oxford University Press, Oxford, UK
17. Aubin J-P (1980) Approximation of elliptic boundary-value problems. Robert E. Kreiger Publishing Company, New York
18. Fletcher CAJ (1984) Computational Galerkin methods. Springer-Verlag, Berlin
19. Gottlieb D, Orszag SA (1977) Numerical analysis of spectral methods: theory and applications. Society for Industrial and Applied Mathematics, Philadelphia, PA
20. Boyd JP (1989) Chebyshev and Fourier spectral methods. Springer-Verlag, Berlin
21. Boyd JP (2001) Chebyshev and Fourier spectral methods, 2nd edn. Dover Publications, New York, NY
22. Boyd JP (1988) Spectral methods in fluid dynamics. Springer-Verlag, Berlin
23. Delic G (1987) Spectral function methods for nonlinear diffusion equations. J Math Phys 28:39–59
24. Herivel J (1975) Joseph Fourier the man and the physicist. Oxford University Press, Oxford, UK
25. Daintith J, Mitchell S, Tootill E (1975) Chambers biographical encyclopedia of scientists. W and R Chambers, Edinburgh, UK
26. Bateman H (1964) Partial differential equations of mathematical physics. Cambridge University Press, Cambridge, UK
27. Jeffreys H, Swirles B (Lady Jeffreys) (1962) Methods of mathematical physics. Cambridge University Press, Cambridge, UK

28. Ascher UM, Mattheij RMM, Russell RD (1995) Numerical solution of boundary value problems for ordinary differential equations. Society for Industrial and Applied Mathematics, Philadelphia, PA
29. Bender CM, Orszag SA (1978) Advanced Mathematical Methods for Scientists and Engineers. McGraw Hill, New York, NY
30. Burrage K (1995) Parallel and sequential methods for ordinary differential equations. Oxford University Press, Oxford, UK
31. Ortega JM, Voigt RG (1985) Solution of partial differential equations on vector and parallel computers. Society for Industrial and Applied Mathematics, Philadelphia, PA
32. Hairer E, Norsett SP, Wanner G (1987) Solving ordinary differential equations I nonstiff problems. Springer-Verlag, Berlin
33. Huebner KH, Thornton EA (1967) The finite element method for engineers, 2nd edn. John Wiley and Sons, New York
34. Kelly LG (1967) Handbook of numerical methods and applications. Addison-Wesley Publishing Company, Reading, MA
35. Atkinson K (1985) Elementary numerical analysis. John Wiley and Sons, New York
36. Butcher JC (2016) Numerical methods for ordinary differential equations, 3rd edn. John Wiley and Sons, New York
37. Hairer E, Wanner G (1996) Solving ordinary differential equations II stiff and differential-algebraic problems. Springer-Verlag, Berlin
38. Lapidus L, Seinfeld JH (1971) Numerical solution of ordinary differential equations. Academic Press, New York, NY
39. Butcher JC (1964) On Runge-Kutta processes of higher order. J Australian Math Soc 4, Part2:179–194
40. Delic G, Malherbe SM (1988) Subroutines for the integration of systems of first order ode's. Comput Phys Commun 48:293–304
41. Ascher UM, Petzold LR (1995) Computer methods for ordinary differential equations and differential-algebraic equations. Society for Industrial and Applied Mathematics, Philadelphia, PA
42. Gear CW (1971) Numerical initial value problems in ordinary differential equations. Prentice-Hall Inc, Englewood Cliffs, NJ
43. Gear CW (2007) Backward differentiation formulas. Scholarpedia 2(8):3162. revision #91024
44. Davis TA (2006) Direct methods for sparse linear systems. Society for Industrial and Applied Mathematics, Philadelphia, PA
45. Kantorovich LV, Akilov GP (1982) Functional analysis, 2nd edn. Pergamon Press, New York, NY
46. Rawitscher GH, Delic G (1986) Sturmian eigenvalue equations with a Bessel function basis. J Math Phys 27:816–823
47. Luke YL (1969) The special functions and their approximation, vol I and II. Academic Press, New York
48. Rektorys K (1982) The method of discretization in time. D. Reidel Publishing Company, Dordrecht, Holland
49. Clenshaw CW (1957) The numerical solution of linear differential equations in Chebyshev series. Proc Camb Philos Soc 53:134–149
50. Elliot D (1960) The numerical solution of integral equations using Chebyshev polynomials. J Aust Math Soc 1:344
51. Fox L, Parker IB (1968) Chebyshev polynomials in numerical analysis. Oxford University Press, Oxford, UK
52. Nettel S (1992) Wave physics: oscillations-solitons-chaos. Springer-Verlag, Berlin
53. Morse PM, Feshbach H (1953) Methods of theoretical physics. Part I. McGraw Hill, New York
54. Kamke E (1971) Differentialgleichungen Lösungsmethoden und Lösungen, vol 1. Gewöhnliche Differentialgleichungen. Chelsea Publishing Company, New York, NY

55. Curtis CF, Hirschfelder JO (1952) Integration of of stiff equations. Proc Natl Acad Sci 38:235–243

Direct Search Optimization Methods

10.1 Why Direct Search Methods?

An optimization algorithm seeks the minimal value of a function of N variables. There are various algorithms [1–9] for such a search and some of these calculate the derivative of the function with respect to these variables during each iteration of the search. Such gradient methods require the numerical calculation of the derivative and may encounter loss of precision as the minimum of the function is approached. An alternative is to apply a direct search algorithm. This is termed "direct search" because only the successive values of the function to be minimized are inspected. Two such methods are described here, as they were applied by the author in two different codes [10–12]. In each case the algorithm, by varying multiple parameters, attempts to improve the fit of a theoretical prediction to experimental data in a non-linear optimization problem by minimizing a function of the form

$$\Delta = \sum_{i=1}^{N} \left[\frac{\left(\frac{d\sigma}{d\Omega}\right)_i^{theory} - \left(\frac{d\sigma}{d\Omega}\right)_i^{experiment}}{\delta\left(\frac{d\sigma}{d\Omega}\right)_i^{experiment}} \right]^2 . \quad (10.1)$$

This is a least squares estimate over points $i = 1, \ldots, N$, of the difference between the theoretical prediction of a model calculation and the measured experimental value at discrete angles. The denominator is the experimental error assigned at each point. The model calculation varies up to six parameters in a search for the minimum in $\Delta(x)$, where x is a vector with six components. This is more than a simple least squares fitting problem [13] as the theoretical predicted values depend on the model

---

Supplementary Information The online version contains supplementary material available at https://doi.org/10.1007/978-3-031-90178-2_10

parameters in a non-linear way and computation of these function values is the most expensive part of the search method. For the minimization search of (10.1) to be successful, the density of the measured values over the angular range must be sufficient to resolve any observed features and an example of a successful search result is shown in Figs. 1 and 2 of [12]. An additional issue is the existence of multiple local minima of the function to be minimized, as these lead to the search stalling when it finds one. Therefore multiple searches need to be initiated from different starting values in an exhaustive grid of the full parameter space, usually with some guidance of expected (realistic) ranges for each parameter.

10.2 Hooke and Jeeves Algorithm

The Hooke and Jeeves (HJ) algorithm [6] searches the N-dimensional space of search parameters spanned by a function $F(x_1, \ldots, x_N)$. It starts from a base point B_0, and performs an exploratory trial move with each parameter, x_i, incremented in turn. This pattern move generates a sequence of points, P_j. An inspection is made for each move to determine if it is a "success", i.e., the function value is lower. If not, the step is reversed and when all N directions have been investigated, then the exploration is complete. The new point is selected as the next base point B_1. A pattern move is then made along the direction pointing from the first to second base point and the corresponding function values are compared to see if it has decreased. If yes, then a new exploration is performed. Even if not, the new base point is still retained as the origin of further exploration to find a smaller value of the function indicating a downward contour. Eventually when there is no reduction in the function value, then either a minimum is reached, or the search has stalled. If the latter is the case, the search is restarted with reduced step sizes. Convergence is assumed when the exploratory steps are reduced below a specified limit. A more detailed explanation is given in Sect. 2.5 of [4]. Thus the HJ algorithm has two important control parameters: the size of the increment in the exploratory moves and the distance between successive increments.

For a function of two variables $f(x_1, x_2)$, Table 10.1 shows cases 1 (a simple quasi-circular function), 2 [7,14] and 3 [7], and case 4 [5]. These are applied in the Fortran code HJ and in EXAMPLES 10.1 to 10.4, where search results of the code are used to show the contours and progress of the search. Table 10.2 also has cases 5 and 6 [5] for functions of three variables where the HJ search algorithm performs well. A final example in Table 10.3 for a four parameter search from [1,14] is incomplete and requires multiple restarts to succeed. These latter three cases are shown in EXAMPLES 10.5 to 10.7. All these EXAMPLES read input files generated by the Fortran code HJ.

Table 10.4 summarizes the progress with statistics for the number of base moves, trial moves, and cost in function evaluations for each of the seven cases.

10.2 Hooke and Jeeves Algorithm

Table 10.1 Cases 1 to 4

Case	$f(x_1, x_2)$	Minima
1	$(1-x_1)^2 + (1-x_2)^2$	$x_1 = x_2 = 1$
2	$100(x_2 - x_1^2)^2 + (1-x_1)^2$	$x_1 = x_2 = 1$
3	$(x_1 + 2x_2 - 7)^2 + (2x_1 + x_2 - 5)^2$	$x_1 = 1, x_2 = 3$
4	$9x_1^2 + 4(x_1 x_2 + x_2^2) - 18x_1 - 4x_2 + 9$	$x_1 = 1, x_2 = 0$

Table 10.2 Cases 5 and 6

Case	$f(x_1, x_2, x_3)$	Minima
5	$(x_1 - 1)^2 + 2(x_2 - 1)^2 + (x_3 - 1)^2$	$x_1 = x_2 = x_3 = 1$
6	$2x_1^2 + 2x_1 x_2 + 3x_2^2$ $+ 8x_2 x_3 + 9x_3^2 + 2x_1 x_3$ $+ 8x_1 - 10x_2 - 30x_3 + 42$	$x_1 = -3$ $x_2 = 0$ $x_3 = 2$

Table 10.3 Case 7

Case	$f(x_1, x_2, x_3, x_4)$	Minimum
7	$(x_1 + 10x_2)^2$ $+ 5(x_3 - x_4)^2$ $+ (x_2 - 2x_3)^4$ $+ 10(x_1 - x_4)^4$	$x_1 = 0$ $x_2 = 0$ $x_3 = 0$ $x_4 = 0$

Table 10.4 Statistics

Case	Base moves	Trial moves	Function calls
1	59	238	308
2	109	395	517
3	115	302	429
4	85	249	346
5	48	150	217
6	179	898	1095
7	131	913	1061

EXAMPLES 10.1–10.7: CASE1 to CASE7
FORTRAN CODE HJ

10.3 Davies, Swann, and Campey Algorithm

The Davies, Swann, and Campey (DSC) method [4,15] uses a set of orthonormal directions that is rotated as the search proceeds (unlike the HJ method where search directions are fixed). The initial directions are the coordinates and a unimodal search is performed in each direction until a minimum is bracketed. After all directions have been so searched, a new direction vector is chosen along the resultant of the unimodal searches and the other directions are obtained by a Gram-Schmidt orthonormalization procedure, to generate the remaining vectors using the resultant as the first one. The directions for which no progress was made are retained for the next iteration of unimodal searches along the new set of orthogonal vectors. When the distance moved in an iteration is less than a prescribed step size, the step size is reduced. When the step size is less than a preset limit, convergence is assumed. In general the DSC method is more efficient than the HJ or Rosenbrock [7] methods. The DSC method may generate undesirable zig-zag searches and a modification has been proposed (but not implemented here) [16]. Two innovations introduced in [11,12] were a Fibonacci [17] linear search [3] and the numerical algorithm of Powell for the Gram-Schmidt orthonormalization [18] described below.

10.3.1 The Fibonacci Search

It can be shown [4,5,17] that for a unimodal function no technique can be guaranteed to find the minimum in less function evaluations than the Fibonacci search that uses the Fibonacci[1] numbers 0, 1, 1, 2, 3, 5, 8, 13, 21, 34, 55, ...

First a definition: a continuous, twice differentiable function is said to be unimodal in a range $a \leq x \leq b$, if there is only one root of $f'(x) = 0$ in that range. Let this root be at $x = x_0$, and correspond to a minimum of $f(x)$. Choose any two values $x_1, x_2,$ of $x \in [a, b]$, if

$$x_1 < x_2 \text{ and } f(x_1) > f(x_2), \text{ then } x_1 < x_0, \qquad (10.2)$$

similarly, if

$$x_1 < x_2 \text{ and } f(x_2) > f(x_1), \text{ then } x_2 > x_0. \qquad (10.3)$$

The unimodal search chooses n internal points on the interval $[a, b]$ where n is one of the values in column 1 of Table 10.5 from [5] and function values are evaluated at the pair of points in the second column. The testing of pairs in (10.2), (10.3) proceeds by successively rejecting the interval where a minimum is not found and adding only one function evaluation in each step, with the cumulative total of function values shown in the third column of Table 10.5.

---
[1] Leonardo Fibonacci, Italian mathematician 1170–1250.

Table 10.5 Fibonacci search

Internal points	Points for first function evaluation	Total number of function values
2	1,2	2
4	2,3	3
7	3,5	4
12	5,8	5
20	8,13	6
33	13,21	7
54	21,34	8
88	34,55	9

10.3.2 Gram-Schmidt Orthonormalization

Previously, in Sect. 9.4.4 a modified Gram-Schmidt (MGS) method was applied to a set of functions, whereas here it is applied to coordinates in an N-dimensional space $\mathbf{E}^N$. If $e^{(i)}, i = 1, \ldots, N, e^{(i)} \in \mathbf{E}^N$, is a set of orthonormal unit vectors that are initially parallel to the axes $x^{(0)} = x_i$, then a linear search is implicit in the notation

$$x^{(k+1)} = x^{(k)} + \lambda e^{(i)}, \ k = 1, \ldots, N, \tag{10.4}$$

where λ is chosen to minimize the function such that

$$\min_{\lambda} f(x^{(k)} + \lambda e^{(i)}). \tag{10.5}$$

Powell in Sect. 2 of [18] has proposed a numerically stable algorithm for the MGS method that starts with N orthonormal direction vectors $\{d_i\}_1^N \in \mathbf{E}^N$. If the search takes a step of length α_i along d_i, then the total displacement is

$$d_0 = \sum_{i=1}^{N} \alpha_i d_i. \tag{10.6}$$

A new orthonormal set of direction vectors $\{d_k^*\}_1^N$ is chosen such that, for $k = 1, 2, \ldots, N$, d_k^* is a linear combination of $d_0, d_1, \ldots, d_{k-1}$.

The MGS algorithm is described as follows. First inspect the sequence of multipliers $\{\alpha_k\}_1^N$ and let α_k be the last non-zero multiplier of the sequence. If $k < N$, then define

$$d_j^* = d_j, \ j = k+1, k+2, \ldots, N, \tag{10.7}$$

and, to begin the iteration that calculates $\{d_j^*\}_2^k$, set

$$t = k,$$
$$s = \alpha_k^2,$$
$$\sigma = \alpha_k d_k. \tag{10.8}$$

The index t is used to count the iteration until $t = 1$, when it is terminated. If $t > 1$, then d_t^* is computed from

$$d_t^* = \frac{(sd_{t-1} - \alpha_{t-1}\sigma)}{\left[s(s + \alpha_{t-1}^2)\right]^{\frac{1}{2}}}. \tag{10.9}$$

Before starting the next iteration, decrease t by one, add α_t^2 to s, add $\alpha_t d_t$ to σ, where now the subscripts have the new value of t and finally set

$$d_1^* = \frac{\sigma}{\sqrt{s}}. \tag{10.10}$$

Stability of this algorithm is discussed in Sect. 3 of [18].

The accompanying Fortran code DSC and examples apply the DSC algorithm described above.

EXAMPLES 10.8 to 10.11: CASE1, CASE5–CASE7
FORTRAN CODE DSC

10.4 Comparison of the Algorithms

Both HJ and DSC direct search algorithms performed well in the real-world applications of [11,12], but skill in their use required some training on how to avoid them stalling. Both methods have input parameters such as starting values, scaling factors, and convergence tolerances that should be considered as variables to adjust when the search stalls. Therefore experimentation with these variables in implementation is always an asset in finding an acceptable function minimum. While the HJ method showed regular progress, the DSC algorithm is prone to stalling and requires numerous restarts. Table 10.6 summarizes a comparison of the number of function values required in EXAMPLES 10.1 to 10.11. The DSC method is not complete for cases 6 and 7 and uses different starting values in multiple restarts. Table 10.7 compares accuracy of convergence for the comparison of the final argument values with HJ and DSC methods.

Table 10.6 Function evaluations

Case	HJ	DSC	HJ/DSC
1	308	74	4.2
2	517	–	-
3	429	-	-
4	346	-	-
5	217	8	27
6	1095	320	3.4
7	1061	1022	1

Table 10.7 Results

Case	HJ	DSC	Exact
1	$x_1 = 1.001$	$x_1 = 1.000$	$x_1 = 1$
	$x_2 = 1.001$	$x_2 = 1.000$	$x_2 = 1$
5	$x_1 = 0.999$	$x_1 = -0.99$	$x_1 = 1$
	$x_2 = 0.999$	$x_2 = 0.99$	$x_2 = 1$
	$x_3 = 0.999$	$x_3 = 0.99$	$x_3 = 1$
6	$x_1 = -3.001$	$x_1 = -3.03$	$x_1 = -3$
	$x_2 = 0.000$	$x_2 = 0.083$	$x_2 = 0$
	$x_3 = 2.000$	$x_3 = 1.948$	$x_3 = 2$
7	$x_1 = 0.01$	$x_1 = 0.165$	$x_1 = 0$
	$x_2 = -0.001$	$x_2 = -0.016$	$x_2 = 0$
	$x_3 = 0.003$	$x_3 = 0.096$	$x_3 = 0$
	$x_4 = 0.003$	$x_4 = 0.096$	$x_4 = 0$

10.5 Programming Problems

1. Apply the Fortran code HJ to cases 6 and 7 with a grid of starting values, to discover if a more efficient search result is possible.
2. Apply the Fortran code DSC to cases 6 and 7 with a grid of starting values, to discover if a more efficient search result is possible.
3. Analyze your results of problems 1 and 2 and compare the relative success and expense of both methods.

References

1. Fletcher R (1965) Function minimization without evaluating derivatives - a review. Comput J 8:33–41
2. Box MJ (1966) A comparison of several current optimization methods, and the use of transformations in constrained problems. Comput J 9:66–77
3. Box MJ, Davies D, Swann WH (1969) Non-linear optimization techniques. ICI Monograph 5. Oliver and Boyd, Ediburgh, UK
4. Murray W (ed) (1972) Numerical methods for unconstrained optimization. Academic Press, New York, NY
5. Dixon LCW (1972) Nonlinear optimization. The English Universities Press Ltd., London, UK
6. Hooke R, Jeeves TA (1961) Direct search solutions of numerical and statistical problems. J Ass Comput Mach 8:212–229
7. Rosenbrock HH (1960) An automatic method for finding the greatest of least value of a function. Comput J 3:175–184
8. Kelley CT (1999) Iterative methods for optimization. Society for Industrial and Applied Mathematics, Philadelphia, PA
9. Powell MJD (1964) An efficient method for finding the minimum of a function of several variables without calculating derivatives. Comput J 7:155–162
10. Delic G (1974) Optical model parameter searches for 16o+11b elastic scattering. Phys Lett 49B:412–414
11. Delic G (1975) Optical model analysis of n+c and c+c elastic scattering. Phys Rev Lett 34:1468–1471
12. Delic G (1990) Optical model analysis of 14n+12c and 12c+12c elastic scattering in the range ecm=35 to 63 mev. Phys Rev 41:2032–2055
13. Björck Å (1996) Numerical methods for least squares problems. Society for Industrial and Applied Mathematics, Philadelphia, PA
14. Powell MJD (1962) An iterative method for finding stationary values of a function of several variables. Comput J 5:147–151
15. Swann WH (1969) A survey of non-linear optimization techniques. FEBS Lett 2(Supplement):S39–S55
16. Hoshino S (1971) On davies, swann and campey minimisation process. Comput J 14:426–427
17. Kiefer J (1957) Optimum sequential search and approximation methods under minimum regularity assumptions. J Soc Ind Appl Math 5(3):105–136
18. Powell MJD (1968) On the calculation of orthogonal vectors. Comput J 11:302–304

Index

B
Basis functions, 17
  Chebyshev, 22
  Fourier, 27
  Legendre, 22
  monomial, 18
  polynomial, 22
Bessel function, 15
  as a power series, 16, 30
  modified first kind, 26
    as a series, 26

C
Chebyshev series for
  associated Legendre polynomials, 36
  Bessel functions, 32
  parabolic cylinder functions, 39
  the derivative, 47
  the integral, 48
Convergence
  Richardson extrapolation, 63, 171
Convolution
  Chebyshev series products, 41
  Fourier series products, 42

D
Divided differences, 57
  and error estimates, 61
  and function derivatives, 61
  and Newton Interpolation, 57
  table, 58

E
Eigenvalue problem, 108, 119, 120
Error
  and convergence rate, 45
  and convergence types, 46
  bound, 44
  in computer arithmetic, 9
  in polynomial interpolation, 61
  loss of significance, 9
  propagation, 9
  truncation, 8, 44

F
Finite differences, 65
  and Newton interpolation, 67
  backward, 65
  central, 65
  difference operator, 68
  forward, 65
  Stirling's central difference formula, 70
  Stirling's formula application, 72
  Stirling's formula coefficients for derivatives, 79
  Stirling's formula error, 79
  tables, 65
Fortran, 10
  compiler, xvii, 11
  intrinsic functions, 11
  numerical models, 11
  standard, 10
Function approximation, 15, 85
  convergent sequence, 86
  Hilbert space
    axioms, 91
    Bessel's inequality, 95
    best approximation, 96
    Cauchy inequality, 91
    inner product, 91
    orthogonality, 93

orthogonal sets, 93
orthonormal sets, 93
orthoprojectors, 94
projection, 96
subspaces, 95
truncated expansions, 94
L2 error, 98
  applications, 102
  Chebyshev series, 99
  Fourier series, 99
  Legendre series, 98
limit of the sequence, 86
metric spaces, 85
  axioms, 86
normed spaces, 88
  axioms, 88
  convergence in norm, 88
  examples, 89
  Minkowski's inequality, 90
  size of elements, 88

G
Gamma function, 16, 30
  of integer order, 26

I
IEEE Standard, 9
Index, 241
Interpolation, 53
  with polynomials, 53
  and derivative, 62
  existence theorem, 54
  Lagrange form, 59
  Newton form, 55

J
Julia, xvii

L
Logical operations, 4

M
Machine representation
  exponent range, 7
  in single precision, 7
  number storage, 8
MacLaurin series, 19

N
Number system
  and machine word length, 7

binary, 3
change of base, 6
decimal, 2
floating point numbers, 7
hexadecimal, 5
machine representation, 7
octal, 5
Numerical
  model, 3
  precision, 11, 176
Numerical computation, 9
Numerical integration
  1-D, 165
    Chebyshev formulas, 182
    definition, 165
    error, 166, 169, 170, 173, 175, 178, 179
    Gauss-Legendre coefficients, 181
    Gauss-Legendre formulas, 176
    higher order rules, 175
    oscillating weight function, 183
    Riemann integral, 166
    Simpson's rule, 172
    trapezoidal rule, 168
  2-D, 189
    definition, 189
    Gauss-Legendre quadrature, 191
    minimal point formulas, 191
    Monte Carlo method, 192
    Simpson's rule, 190
Numerical solution of ODEs, 197
  background and definition, 197
  diffusion and wave equation, 198
  on a physical grid, 198
    Euler's method, 199
    multi-step methods, 199
    Predictor-Corrector, 204
    Runge-Kutta, 200
    Stiff ODEs, 205
  second-order ODEs, 224

O
Operator equations, 107
  approximation, 108
  determinants, 116
  linear algebra libraries, 150
    BLAS, 150
    LAPACK, 150
    LINPACK, 150
  matrices, 110
  matrix iteration methods, 140
    condition number, 139

convergence criteria, 143
factorization, 140
Gauss-Seidel iteration, 142
Jacobi iteration, 141
rate of convergence, 146
relaxation, 146
matrix operators, 118
matrix solution methods, 120
Cholesky factorization, 130
condition number and stability, 135
Crout factorization, 129
Doolittle's method, 128
Gaussian elimination, 121
Gauss-Jordan elimination, 132
inverse matrix, 125
LU decomposition, 126
matrix norms, 135
partial pivoting, 124
tridiagonal matrices, 130
permanent, 117
sparse matrices, 114
vectors, 109
Optimization methods, 233
direct search algorithms, 233
comparison of algorithms, 238
Davies, Swann, and Campey, 236
Fibonacci search, 236
Gram-Schmidt orthonormalization, 237

Hooke and Jeeves, 234
Orthogonality, 24
Chebyshev polynomials, 24
Legendre polynomials, 24
trigonometric functions, 28
Orthonormalization
Gram-Schmidt method, 211, 236, 237

P
Polynomials, 22
Chebyshev
zeros, 23
Chebyshev series, 25
Legendre
zeros, 184
Legendre series, 24
Preliminaries
About the Author, xxvii
About this Book, xvii
Acknowledgement, xv
Copyright, xix

Foreword, xi
Fortran examples, xxi
Permissions
Table 1, xx
Preface, xiii

R
Recurrence
Associated Legendre polynomials, 36
Bessel functions, 16, 31
Chebyshev polynomials, 24
Chebyshev series
evaluation, 26
Legendre polynomials, 23
parabolic cylinder functions, 39
stability, 16
trigonometric functions, 29
trigonometric series, 29
Roots of functions, 153
definition, 153
ill-posed problems, 161
intermediate value theorem, 154
Legendre polynomial, 162
numerical methods, 154
bisection, 154
comparison, 160
convergence, 154, 155, 158, 160
Newton, 156
secant, 159
spherical Bessel function, 162

S
Spectral function method, 208
choice of basis, 208
incorporating boundary conditions, 209
non-linear diffusion equation, 216
orthonormalization, 211
second-order spatial derivative, 212
spectral mappings, 214
Spectral function method (double), 217
non-linear diffusion equation
by recurrence, 220
by time stepping, 218

T
Taylor
polynomial, 20
series, 19, 157, 170, 174
error term, 20

GPSR Compliance
The European Union's (EU) General Product Safety Regulation (GPSR) is a set of rules that requires consumer products to be safe and our obligations to ensure this.

If you have any concerns about our products, you can contact us on

ProductSafety@springernature.com

In case Publisher is established outside the EU, the EU authorized representative is:

Springer Nature Customer Service Center GmbH
Europaplatz 3
69115 Heidelberg, Germany